Edible Roots and Underground Stems of Ethnic India

Edible Roots and Underground Stems of Ethnic India

S.K. Sood & Ved Prakash
Department of Biosciences
Himachal Pradesh University, Shimla - 171005

SATISH SERIAL PUBLISHING HOUSE
403, Express Tower, Commercial Complex, Azadpur, Delhi - 110033 (INDIA)
Phone : 011-27672852 Fax : 91-11-27672046
E-mail : info@satishserial.com, hkjain1975@yahoo.com
Website : www.satishserial.com

Published by :

SATISH SERIAL PUBLISHING HOUSE
403, Express Tower, Commercial Complex, Azadpur, Delhi - 110033 (INDIA)
Phone : 011-27672852 Fax : 91-11-27672046
E-mail : info@satishserial.com, hkjain1975@yahoo.com

ISBN 81-89304-26-7

Printed at
SALASAR IMAGING SYSTEMS, DELHI-35

PProf. T.N. Lakhanpal
M.Sc.(Hons.), Ph.D.,
F.P.S.I., F.P. S.M.
Ex. Head and Dean
Faculty of Life Sciences, and
Director, Institute of Integrated
Himalayan Studies

Department of Bio-Sciences
Himachal Pradesh University,
Shimla-171005. India
Phone 0177-230946 (O)
0177-2621229,2625456 (R)
FAX- 0177-2621229
E-mail - tezlakhanpal@rediffmail.com

Foreword

It is a matter of pleasure and honor to write a foreword for the compendium"Edible Roots and Underground Stems of Ethnic India"authored by Prof. S.K.Sood and Mr. Ved Parkash.

In the new millenium when the population is increasing dispropationately to the already utilizable resources, there is a need to explore new sources of food and feed as well as exploit these which have been utilized by human being traditionally. Out of a few hundred thousand species which occur in this world, only about two dozen are being exploited for meeting the food demands of the population. Majority of the plants used as food are in the form of their seeds or grains. There is a large number of plants in which underground parts are known to provide an equally rich source of food for human consumption. In Indian tradition, non cereal food is recommended during many rituals especially fasts, feasts and festivals. These types of plants are listed by different workers in the lists of edible plants. There is no separate consolidated information available on these. The present compendium fills this gap and provides consolidated information at one place on the edible roots and underground stems of Ethnic India. The species are listed alphabetically and each species provides information on distribution and uses of the parts along with scientific name, common name, family, folk uses and active ingredients.

The compendium is divided into five chapters preceded by Preface and abbreviations. Chapters I -IV give an overview of the importance of plants for human welfare, richness of Indian flora, overview of the previous work and the arrangement of data and its presentation, respectively. Chapter V describes the edible roots and underground parts in detail, providing all information on all aspects of science and art of use. It includes 401 plant species belonging to 186 genera and 67 families. The plant species belong to Pteridophytes, Gymnosperms and Angiosperms. It also provides information on the nutritive value and ancillary uses. The work is documented with coloured photographs of important plant species and all the relevant literature is properly listed. Very useful information has been gathered and consolidated within this volume. It is expected to generate interest in people working in all spheres of Biology.

(T.N. LAKHANPAL)

Preface

The plants of our planet are of paramount importance to our continued existence. In the scenerio of expanding humanity and depleting resources, it is a challenging task for the policy-makers to provide sustenance to over 500 million people suffering from malnutrition and about 20 million people starving to death annually due to hunger and poverty, wild edible plants are the only ray of hope. In India (unique in the richness of its plant wealth) more than 3,000 wild plant species constitute the major food component of our 550 tribal communities under 227 ethnic groups. And these are used particularly during periods of food scarcity and outbreak of famine or starvation. A few books have been done globally on wild edible plants but compilation on "Edible Roots and Underground Stems" i.e. plants eaten or otherwise consumed by Indians, was badly needed. To fill this gap, the present compendium has been written that affords both laymen and specialist alike comprehensive encyclopaedia of data from disparate sources in an easy-to-read understand style. The information on the usage of 401 plant species belonging to 186 genera under 67 families has been presented. To make the compilation useful, correct nomenclature of each species with complete citation, synonyms (if any), names in Indian languages and English, distribution, flowering and fruiting period and chemical constituents are given. Not only this, eight useful appendices have also been included. It is hoped that this compendium will be useful to students, researchers, teachers and scientists in traditional and agricultural universities, besides research institutions on plant breeding, economic botany, crop improvement and plant genetic resources.

This venture could not have been possible without the valuable advice and incessant encouragement of Prof. L.R. Verma, Vice-Chancellor, Himachal Pradesh University, Shimla, for which we express our sincere gratitude to him. Well deserved appreciation is due to Prof. T.N. Lakhanpal, Prof. D.C. Kalia and other colleagues of Department of Biosciences for their co-operation and well wishes; to Mr Suresh, Ms Anjna and Mr Munish for their ungrudging help in critically reading the various drafts of this manuscript, and to Mr Bal Krishan for his help in various ways. Thanks are also due to authors of books, papers and figures listed in the compendium. Finally, we would like to thank the members of our families for moral support and unflinching patience during the course of the work.

Suggestions for further improvement of the compendium will always be welcomed and entertained wholeheartedly.

S.K. Sood, Ved Prakash

Acknowledgements

I would acknowledge with utmost sincerity and earnestness that but for the intensive and immense guidance and direction of my Research Supervisor, Dr. S.K Sood, Professor and Chairman, Department of Bio-Sciences, Himachal Pradesh University, Shimla, India without which the present work would not have been what it happened to be.

I take this opportunity to extend my gratitude and kind regard to entire faculty members of Department of Bio-Sciences for their help and encouragement.

Grateful thanks are also extended to the library staff of Himachal Pradesh University, Shimla, for their kind help and co-operation.

I can neither compensate nor reimburse the debt of the everlasting affection, blessing and sacrifices of my parents who, untiringly not caring for their welfare, always rendered assistance with a smile and without a wrinkle on their faces to enable me to do my research work with added enthusiasm and earnestness.

I also thank my senior lab-mates Mr. Dheeraj, Ms. Anjana, Mr. Suresh, Mr. Munish, for providing a stimulating working environment in the laboratory and their kind help in various ways.

My thanks are due to score of friends especially Anil, Shiv, Hemraj, Radhika, and well wishers who boosted me up and helped me a lot in many ways. Specially, I want to thank two of my best friends Mr. Kashmir and Ms. Romita for encouraging and helping me during entire work and all the tough times that relieved me of undue pressure of work.

Love and inspiration of my brother, Khushal and sisters, Ranju and Babli is greatly acknowledged.

My thanks are also due to Mr. Jitender Rana and Mr. Ghanshyam Lal for their help in photography. I am also thankful to my senior Mr. Murari for assistance in typing and setting the entire manuscript and Mr. Tilak Sharma for printing and binding proficiently.

Last but not the least I owe my gratitude to the almighty God and my two local deities 'Shikari Devi' and 'Dev Kamarunag' whose benediction and benison are always with me since my birth without which I would never been able to complete this work.

Ved Prakash

Abbreviations

α	Alpha	Mal.	Malayalam
&	And	Man.	Manipuri
Ass.	Assam	Mar.	Marathi
Beng.	Bengali	Megh.	Meghalaya
β	Beta	m	Meter
Bomb.	Bombay	Me	Methyl
Ca	Calcium	μ	Micro
C.	Central	mg	Milligram
Cl	Chlorine	N.	North / Northern
δ	Delta	N.E.	North East
E.	East / Eastern	N.W.	North West
Eng.	English	Ory.	Oriya
et al.	et alia (and other authors)	%	Per cent
etc.	et cetera (so on)	P	Phosphorus
γ	Gamma	Pub.	Punjabi
Garh.	Garhwal	Sant.	Santhali
Guj.	Gujarati	Sans.	Sanskrit
H.P.	Himachal Pradesh	Si	Silicon
i.e.	id est. (that is)	Sp.	Specie
I	Iodine	Spp.	Species
I.U.	International Units	S	Sulphur
Fe	Iron	Syn.	Synonyms
J. & K.	Jammu and Kashmir	Tam.	Tamil
Kan.	Kannada	U.P.	Uttar Pradesh
Ker.	Kerala.	var.	Variety
Kon.	Konkani	vit.	Vitamin
Kum.	Kumaon	W.	West / Western
M.P.	Madhya Pradesh		
Mg	Magnesium		

CONTENTS

Introduction

Man and Plants

Plants have existed on land for the last 430 million years. Before the start of organized agriculture, the food needs of prehistoric nomads was met largely from forests. There are about 2,20,000 species of the flowering plants all over the world and the very survival of humanity is dependent on them for their food, fuel, fodder and medicine and most of their material requirements. Tropical countries are very large in floristic diversity. In India, approximately 80 - 90 per cent of the species of the flowering plants occur in the forest area, and northeastern part is the richest in the floristic diversity. About 3,000 species all over the world are being commonly used in agriculture, horticulture and forestry of which 30 species provide most of the food requirement of man. The utility of the all the wild plants for various purposes was not recognized in one day or in one century, it is the result of progressive development of human cultures (Mittre, 1981). Infact, ethnobotany denotes the entire realm of useful relationship between plants and man (Jain, 1986). A vast store of information and knowledge on potentially useful plants is still intact with the tribal communities. The role of ethnobotanical studies in searching new plant sources of food has also been emphasized (Schultes, 1960; Jain, 1987). In overall perspective, tribal knowledge of plants is important not just for their sustenance and livelihood, but for the wider world (Maheshwari, 1987). Unfortunately, this great treasure is now under assault everywhere in the world under the impact of industrialization, and be documented before it is lost forever.

India – Richness of Plant Wealth

India lying between the latitude 8° 04¢ N to 37° 06¢ N and longitude 68° 07¢ E to 97° 25¢ W is unique in the richness of floristic composition, agricultural diversity and ethnic variability. Of the over 15,000 species of higher plants occurring in India, 9, 000 are economically useful; of which 7,500 are medicinal, 3,900 are edible, 700 are culturally important, 525 are used for fodder, 300 for pesticides and insecticides, 300 for gum, resin and dye and 100 provide incense and perfume (Anon., 1994). It is estimated that in India about 1,000 species are consumed as foodplants chiefly by its 53 million native or tribal communities under 227 ethnic groups occupying about 18.74 per cent of the total area under varied geographical and climatic zones of over 5,000 forested villages (Singh & Arora, 1978; Anon., 1994). Based on the plant part(s) eaten, the edible diversity comprises of roots/ tubers (145 spp), leafy vegetables (521 spp), buds/flowers (101 spp), fruits (647 spp) and seeds / nuts (16 spp) (Anon., 1994). According to Nayer *et al.* (1989-1994) about 20 per cent of Indian flowering plants are useful to man directly. These plants provide inexpensive food, thereby extending the very base of our food security system.

Overview of the Work on Edible Plants

Plants are the prime life support system of our biosphere. India's richness in wild edible plant wealth is known since ancient times as the Hindu scriptures contain a plethora of information about the names of the uses of many plants. Traditional agriculture and forest resources are still the primary means of livelihood of a large number of people in the country. As a matter of fact, the wild food plants occupy an important place in the rural dietary habits particularly during the period of food scarcity and outbreak of famine or starvation (Arora, 1991). More importantly, studies on nutritional values of wild plantfoods are of considerable significance as they help to pinpoint traditional food resources of rural and ethnic communities. There are a few major publications on country-wise basis for wild edible plants, namely for Britain (Cameron, 1917; Hill, 1939), Philippines (Wester, 1925), U.S.A. and Canada (Saunders, 1934; Castetter, 1935; Medsger, 1943), West Africa (Dalzell, 1937; Irvine, 1952), Northern Canada (Porsild, 1937), World (Clute, 1943; Tanaka, 1976; Meninger, 1977), Eastern North America (Fernald & Kensy, 1943), Russia (Klesnikov, 1943), China (Portefied, 1951; Cheng, 1965), South Pacific Islands (Barrau, 1959) and India (Singh & Arora, 1978; Kunkel, 1984). In India, the pioneering work on edible plants was that of King (1869) who reported some plants used during the times of famine from the region of Marwar. However, the pioneering work on Indian ethnobotany had been that of Janaki Ammal (1955), who studied subsistence on food plants from certain tribals of South India. The first bibliography of ethnobotany in India was published in the early eighties (Jain *et al.*, 1984), followed by a recent detailed one describing over 1600 references (Jain, 2002). Scrutiny of literature revealed that there has been an upsurge of activities in the last four decades for ethnobotanical pursuits in different phytogeographical areas and tribes of India as well as the literature on wild edible plants is large and scattered (see Table I). Particular mention can be made of works carried out by various workers on wild plantfoods of different tribes like Baster (Jain, 1963; Sahoo, 1996), Purulia (Jain & De, 1964), Northeastern tribals (Arora, 1981), Garos (Biswas & Bhuyan, 1983), Gaddies (Lal *et al.*, 1994), Nagas (Jamir, 1995). Tarie tribals (Bajpayee & Dixit, 1996), Kashmiri tribes (Raveendran, 1996) and Chenchu tribes (Kumar & Pullaiah, 2003). Besides, mention on wild edibles has also been made in various regional floras, books, compendia on economic plants and ethnobotanical studies Gammie, 1902; Anonymous, 1948-1976; Krishnamurthy, 1953; Sundararaj & Balasubramanyam, 1955; Nathawat & Deshpande, 1960; Badhwar & Fernandez, 1964; Lancaster, 1964; Maheshwari & Singh, 1965; Agrawal & Malhotra, 1970; Gaur, 1977; Singh & Arora, 1978; Atkinson, 1982; Kaul *et al.*, 1982, 1985; Parmar & Kaushal, 1982; Singh *et al.*, 1983; Kaul & Singh, 1985; Agarwal, 1986, 2003; Ambasta, 1986; Anonymous 1986a, b; Girach *et al.*, 1988; Negi, 1988; Watt, 1889-1896; Haridasan *et al.*, 1990; Bennet *et al.*, 1991; Singh, 1996; Chandra, 1997; Roy *et al.*, 1998; Samant *et al.*, 2001; Sood *et al.*, 2001; Rana, 2004; Sood & Thakur, 2004; etc. Unfortunately, no consolidated treatise exclusive on edible roots, tubers and rhizomes of ethnic India has appeared so far. The present work was therefore, undertaken with a view to synthesize the current state of knowledge on the subject.

Table 1. Chronological Arrangement of Important Works on Edible Plants of India

S. No.	Author (s)	Aspects Studied
1.	Shortt (1887-1888)	Wild plants and vegetables used as food by people in famine times.
2.	Watt (1889-1896)	Dictionary of economic products of India.
3.	Gammie (1902)	Plants used during famine and scarcity in Bombay Presidency.

S. No.	Author (s)	Aspects Studied
4.	Bal (1942)	Useful plants of Mayurbhanj (Orissa).
5.	Anonymous (1948 – 1976)	Wealth of India.
6.	Dastur (1951)	Useful plants of India.
7.	Lal *et al.* (1952)	Utilization of tender bamboo shoots as food.
8.	Krishnamurthy (1953)	Horticulture and economic plants of Nilgiris.
9.	Nadkarni (1954)	Indian Materia Medica.
10.	Sundararaj & Balasubramanyam (1955)	Economic plants of South India.
11.	Anonymous (1958)	Economic plants of Lucknow District.
12.	Bhargava (1959)	Usual and supplementary food plants of Kumaon.
13.	Vartak (1959 a)	Edible wild plants of Poona district.
14.	Vartak (1959 b)	Edible wild plants of Maharashtra and Goa.
15.	Nathawat & Deshpande (1960)	Economic plants of Rajasthan.
16.	Gupta (1962)	Unusual and interesting food plants of Garhwal Himalaya.
17.	Jain (1963)	Wild plant foods of tribals of Baster.
18.	Badhwar & Fernandez (1964)	Edible wild plants of Himalayas.
19.	Badhwar *et al.* (1964)	Some useful aromatic plants.
20.	Jain & De (1964)	Plant foods among the tribals of Purulia (W. Bengal).
21.	Lancaster (1964)	Food plants of India.
22.	Jamir (1995)	Wild edible plants of Nagaland.
23.	Maheshwari & Singh (1965)	Dictionary of economic plants of India.
24.	Choudhary (1967)	Vegetables.
25.	Gupta & Kanodia (1968)	Famine plants in dry regions of India.
26.	Yawalkar (1969)	Vegetable crops of India.
27.	Agrawal & Malhotra (1970)	Unreported economic plants of India.
28.	Gopalan *et al.* (1971)	Nutritive value of Indian foods.
29.	Pal & Banerjee (1971)	Food plants used by the tribals of Andhra Pradesh and Orissa.

S. No.	Author (s)	Aspects Studied
30.	Sangal (1971)	Food plants used by the tribals of Andaman and Nicobar Islands.
31.	Bhandari (1974, 1979)	Famine food in Rajasthan desert.
32.	Datar & Vartak (1975)	Wild edible plants from Karnal bird sanctuary.
33.	Deb (1975)	Economic plants of Tripura.
34.	Gaur (1977)	Wild edible fruits of Garhwal hills.
35.	Jain et al. (1977)	Wild edible plants in Bazars of Meghalaya.
36.	Raichaudhury *et al.* (1977)	Edible jungle leaves.
37.	Singh & Arora (1978)	Wild edible plants of India.
38.	Saxena (1979)	Plant foods of western Rajasthan.
39.	Atal *et al.* (1980)	Emergency food plants of Jammu and Kashmir state.
40.	Pangtey (1980)	Wild edible fruits of Kumaon hills.
41.	Arora (1981)	Native food plants of northeastern tribals.
42.	Hajra & Chakraborty (1981)	Wild plants sold in the Lal Market of Gangtok.
43.	Singh & Singh (1981)	Wild edible plants of eastern Rajasthan.
44.	Vartak (1981)	Future prospects of wild edible plants from hilly regions of Maharashtra and Goa.
45.	Atkinson (1982)	Economic botany of the Himalayan region.
46.	Kaul *et al.* (1982)	Wild edible plants of Kashmir.
47.	Pangtey *et al.* (1982)	Usual and supplementary food plants of Kumaon.
48.	Parmar & Kaushal (1982)	Wild fruits of the sub-Himalayan region.
49.	Biswas & Bhuyan (1983)	Food plants of Garo Hills, Meghalaya.
50.	Gaur & Semwal (1983)	Some little known wild edibles of Garhwal Himalaya.

S. No.	Author (s)	Aspects Studied
51.	Singh *et al.* (1983)	Dictionary of economic plants of India.
52.	Arora & Nayar (1984)	Wild relatives of crop plants in India.
53.	Manna & Samaddar (1984)	Useful plants in Netarhat plateau of Palamau district, Bihar.
54.	Arora (1985)	Genetic resources of less known cultivated food plants.
55.	Kaul *et al.* (1985)	Wild food plants of Ladakh.
56.	Kaul & Singh (1985)	Wild edible plants of ethnobotanical significance.
57.	Malhotra *et al.* (1985)	Additional notes on the wild edible plants of India.
58.	Nilegaonkar *et al.* (1985)	Nutritional value of some wild food plants from Pune and neighbouring districts
59.	Paroda *et al.* (1985)	Life support plants species, diversity and conservation.
60.	Raichaudhury *et al.* (1985)	Wild edible plants in Calcutta market.
61.	Shankar (1985)	Life support species in Indian Thar desert.
62.	Singh & Mahashwari (1985)	Some wild food plants of the tribals of Varanasi district, (U.P.).
63.	Singh & Singh (1985, 1988)	Some wild edible plants in the markets of Manipur.
64.	Agarwal (1986)	Economic plants of India.
65.	Ambasta (1986)	The useful plants of India.
66.	Anonymous (1986a)	Economic plants of India.
67.	Anonymous (1986b)	The useful plants of India.
68.	Dagar & Dagar (1987)	Useful pteridophytes of Andaman and Nicobar islands.
69.	Jain & Sinha (1987)	Some emergency and supplementary foods among aboriginals in India.
70.	Pandey & Pangtey (1987)	Edible ferns of Kumaon (W. Himalaya).
71.	Uma & Singh (1987)	Edible species of *Diospyros* in India.
72.	Vartak & Kulkarni (1987)	Wild leafy vegetables from hilly regions of Pune.

S. No.	Author (s)	Aspects Studied
73.	Girach *et al.* (1988)	Wild edible plants from tribal pockets of Orissa.
74.	Negi (1988)	Wild edible plants of U.P. hills.
75.	Raju & Krishna (1988, 1990)	Wild edible plants of Sikkim.
76.	Nayer *et al.* (1989-1994)	Economic plants of India (Diet).
77.	Bonde *et al.* (1990)	Nutritional composition and the fruits and Downpalms from the west coast of India.
78.	Haridasan *et al.* (1990)	Wild edible plants of Arunachal Pradesh.
79.	Maniyam & Ramachandran (1990)	Wild edibles in Coimbatore district.
80.	Negi *et al.* (1990)	Use of Seabuckthorn as a pickle plant.
81.	Arora (1991)	Native food plants of northeastern tribals.
82.	Bennet *et al.* (1991)	Food from forest.
83.	Pruthi (1993)	Major spices of India.
84.	Dutt (1993)	Materia Medica of Hindus.
85.	Kulkarni & Kumbhojkar (1994)	Palatable fodder grasses from Pachgaon Parvati area in Pune district.
86.	Lal *et al.* (1994)	Supplementary foods of Gaddies in H.P.
87.	Vartak & Suryanaryana (1995)	Wild edible plants from Susala island, Pune district.
88.	Arora & Pandey (1996)	Wild edible plants of India, diversity conservation and use.
89.	Bajpayee & Dixit (1996)	Foodstuff of tribals of Tarie-region (U.P.).
90.	Borthakur (1996)	Wild edible plants in markets of Assam.
91.	Islam (1996)	Ethnobotany of certain underground parts of N.E. region.
92.	Nazarudeen *et al.* (1996)	Food plants of Paniyar community in north Kerala.
93.	Radhakrishnan *et al.* (1996)	Wild edible plants of Kerala.

S. No.	Author (s)	Aspects Studied
94.	Raveendran (1996)	Some lesser-known edible plants among tribals of Kashmir.
95.	Sahu (1996)	Life support promising food plants among aboriginals of Bastar (M.P.).
96.	Sharma (1996)	Emergency food in Kautilaya's Arthashashlra-an ancient Sanskrit work.
97.	Singh & Viswanathan (1996)	Useful pteridophytes of India.
98.	Singh (1996)	Wild edible plants of Mandi district (N.W. Himalaya).
99.	Sinha (1996)	Wild edible plants of Manipur.
100.	Ansari (1997)	Wild edible plants of Shevory and Kolly hills of south India.
101.	Chandra (1997)	Edible plants of forestry origin.
102.	Samant & Dhar (1997)	Economic potential of wild edible plants of Indian Himalaya.
103.	Dhar *et al.* (1998)	Wild plant resources and use pattern in Askot and wild life sanctuary of Kumaon (W. Himalaya).
104.	Goud *et al.* (1999)	Wild food plants of Kurnool district (A.P.).
105.	Krishna *et al.* (1999)	Food plants of Konda Reddies of East Godavari district (A.P.).
106.	Mehtani & Ingle (1999)	Plants food flavours.
107.	Singh *et al.* (1999)	Wild edible plants of Tripura.
108.	Singh (1999)	Useful plants of Kulu district in north-western Himalayas.
109.	Raju (2000)	Wild plants of Indian subcontinent and their economic use.
110.	Singh & Kumar (2000)	Ethnobotanical wisdom of Gaddies tribes of Western Himalayas.
111.	Aswal (2001)	Zingiberaeae of western Himalaya.
112.	Das & Chattopadhyay (2001)	Nutritional evaluation of edible non-timber forest produce.
113.	Samant *et al.* (2001)	Diversity and distribution of wild edible plants of Indian Himalaya.

S. No.	Author (s)	Aspects Studied
114.	Shukla *et al.* (2001)	Wild edible plants of district Bilaspur (M.P.).
115.	Singh *et al.* (2001a)	Edible wild plants of trans-Himalayan cold desert.
116.	Singh *et al.* (2001b)	Wild edible aquatic plants of Manipur valley.
117.	Sood *et al.* (2001)	Ethnobotany of cold desert tribes of Lahul-Spiti (N.W.) Himalaya.
118.	Lalramnghinglova (2002)	Edible plants of Mizoram.
119.	Patole & Jain (2002)	Wild edible plants of Pachmarhi Biosphere reserve (M.P.).
120.	Agarwal (2003)	Directory of Indian economic plants.
121.	Kulkarni *et al.* (2003)	Nutritional potential of leafy vegetables consumed by Mahadeokoli tribe in W. Maharashtra.
122.	Kumar & Pullaiah (2003)	Wild edible plants of Chenchu tribes of Prakasm district (A.P).
123.	Rana (2004)	Lesser known food plants of H.P.
124.	Sood & Thakur (2004)	Ethnobotany of Rewalsar Himalaya.
125.	Rana *et al.* (2005)	Genetic diversity of Himalaya.

Presentation and Arrangement of Data

For compiling a widely scattered large mass of data on edible roots and underground stems from published ancient treatises, books, monographs, floras, compendia, materia medica and research journals on economic botany and ethnobotany, the standard methodology as outlined by Schultes (1960) and Jain (1964, 1967) was adopted. However, the information presented for English, Hindi, Sanskrit and Regional names, distribution, reproductive cycle and active constituents for each species, wherever known have been complied and collated from Hooker (1872-1897), Nadkarni (1914), Kirtikar & Basu (1984), Anonymous, (1948-1976), Chopra *et al.* (1956, 1968), Jain (1968), Rastogi & Mehrotra (1990, 1991, 1993, 1995), Asolkar *et al.* (1992) and Bhattacharjee (1998). Besides the information on wild edible roots and underground stems, an attempt has also been made presently to include information on plants cultivated by mankind in India.

Over 401 plant species known for their edible roots and underground stems are arranged alphabetically furnishing the botanical name, synonyms if available, family, English Hindi, Sanskrit and Regional names, distribution, reproductive cycle, part(s) used, folk use(s), active constituents, relevant references and remarks, if any. To make reading interesting, the text has been explained with the help of 28 photographs, 4 histograms and 4 tables. Appendices include information on total number of edible genera and species of roots and underground stems under various divisions of plant kingdom (Appendix I), families of edible roots and underground stems of ethnic India (Appendix II), total number of edible species of roots and underground stems under predominant genera (Appendix III) and Index to Botanical names (Appendix IV), Index to English names (Appendix V), Index to Hindi names (Appendix VI), Index to Sanskrit names (Appendix VII) and Index to Regional names (Appendix VIII).

Edible Roots and Underground Stems of Ethnic India

Abelmoschus crinitis Wall.

Syn : *Abelmoschus racemosus* Wall.; *Bamia cancellata* Wall.; *Bamia fusiformis* Wall.; *Hibiscus cancellatus* Roxb.; *H. crinitis* Wall.; *H. racemosus* Lindley.

Family : Malvaceae.

Regional Name	:	*Beng.* : Bankapas, Birkapas.
Distribution	:	N.W. Himalaya in Kumaon, Garhwal region upto 1,400 m and in subhimalayan tract from Kashmir eastwards.
Reproductive Cycle	:	July - January(Next year).
Part Used	:	Tuberous Fusiform Roots.
Folk Uses	:	Tuberous roots eaten raw or cooked as vegetable by the tribals of Kumaon, Garhwal, Manipur and Palamou Hills.
References	:	Arora & Pandey (1996), Bennet *et al.* (1991), Samant *et al.* (2001) Singh (1996).

Abelmoschus moschatus Medic.

Syn : *Hibiscus abelmoschus* Linn.; *H. chinensis* Wall.; *H. flavescens* Cav.; *H. ricinifolius* Wall.; *A. rugosus* Wight & Arn.; *H. sagittifolius* Kurz.; *H. spathaceus* Wall.

Family : Malvaceae.

English, Hindi, Sanskrit and Regional Names:

Eng. : Ambrette plant, Musk mallow;
Hindi : Muskdana, Kasturidana;
Sans. : Latakasturikam;
Beng. : Kasturidana, Muskdana;
Kan. : Kadu Kasturi;
Mal. : Kattu Ksturi;
Mar. : Musk -bhendi;
Tam. : Kattukasturi, Varttilai;
Tel. : Kasturibhenda, Kasturibhend.

Distribution : Deccan foothills of Himalaya. Cultivated in lower foothills of H.P. upto 2,000 m. Hotter parts of India.

Part Used : Roots.

Folk Uses : Roots eaten and considered as good tonic.

References : Ambasta (1986), Bennet (1987), Chandra (1997), Sharma & Malhotra (1984).

Abutilon indicum (Linn.) Sweet.

Syn : *Abutilon asiaticum* W & A. Prodr.

Family : Malvaceae.

English, Hindi, Sanskrit and Regional Names:

Eng. : Indian Mallow;
Hindi : Kangli;
Sans. : Atibala;
Beng. : Potarit;
Tam. : Paniyarattitti;
Tel. : Tuttrurabenda.

Distribution : Kumaon and Uttaranchal region of N.W. Himalaya upto 1,000m. Generally distributed throughout India.

Part Used : Tuberous Roots.

Folk Use : Root tubers edible.

References : Gupta & Kanodia (1968), Rana *et al.* (2005), Samant *et al.* (2001), Saxena (1979), Watt (1889-1896).

Acanthopanax trifoliatum (Linn.) Voss.

Syn : *Acanthopanax aculetum* Linn.

Family : Araliaceae.

Distribution	:	N.E. Himalaya in Assam.
Part Used	:	Roots.
Folk Use	:	Roots cooked and eaten.
References	:	Ambasta (1986), Arora & Pandey (1996).

Aconitum balfourii Stapf.

Family : Ranunculaceae.

Distribution	:	N.W. Himalaya (Kumaon region): 3,500 - 4200 m.
Part Used	:	Roots.
Folk Use	:	Roots eaten raw or boiled.
References	:	Anonymous (1948-1976), Samant *et al.* (2001).
Remarks	:	A poisonous herb to be consumed with care.

Aconitum heterophyllum Wall.

Syn : *Aconitum cordatum* Royle

Family : Ranunculaceae.

Hindi and Sanskrit Names:

Hindi	:	Atis;
Sans.	:	Ativisha.

Distribution	:	N.W. Himalaya (Kumaon): 3,500– 4,200 m.
Part Used	:	Roots.
Folk Use	:	Roots eaten after repeated boiling.
References	:	Anonymous (1948-1976), Samant *et al.* (2001).
Remarks	:	A plant used as fish poison and should be eaten only after proper boiling.

Acorus calamus Linn.

Family : Araceae.

English , Hindi and Sanskrit Names:

Eng. : Sweet-flag;
Hindi : Bach, Ghorabach, Safed-bach;
Sans. : Bhuta-nashini, Galani, Jatila, Ugragandha, Vacha.

Distribution : Damp marshy places in various parts of India from 900-1,800 m. Cultivated in Karnataka and other regions.

Part Used : Rhizomes.

Folk Uses : Rhizomes peeled and washed to remove the acrid elements. The 'calamus oil' obtained from rhizomes has spicy fragrance and used for flavouring alcoholic beverages, and also as spice and condiments.

Active Constituents : Dry rhizome contains asaryl aldehyde, acorin (glucoside), asamyl alcohol, eugenol, asarone, starch, tannins, mucilage and saponins.

References : Anonymous (1999), Asolkar *et al.* (1992), Bhattacharjee (1998), Chevallier (1996), Harborne & Baxter (2001), Jain (1968), Kirtikar & Basu (1984), Purohit (2000), Rastogi & Mehrotra (1990, 1991, 1993, 1995), Sharma & Sood (1997), Watt (1889-1896).

Ajuga macrocarpa Wall. ex Benth.

Family :Lamiaceae.

Distribution : Subtropical Himalaya.

Part Used : Roots.

Folk Uses : Roots used for consumption and also as tonic.

Reference : Ambasta (1986).

Alisma plantago-aquatica Linn.

Syn : *Alisma plantago* Linn.; *Alisma aquatica* Linn.
Family : Alismataceae.

English and Regional Names:

Eng. : Mad dog weed, Water plantain;
Mani. : Ishingy -empa, Kakthrum.

Distribution : Tropical Himalaya. N.E. Himalaya. Wetlands of India.

Reproductive Cycle : November-February (Next year).

Part Used : Rootstocks.

Folk Uses : Rootstocks eaten after drying. The young leaves alongwith the inflorescence and roots used in the preparation of 'Iromba' - a special food made by Manipuri tribals.

References : Ambasta (1986), Bennet *et al.* (1991), Singh *et al.* (2001b), Singh (1996).

Allium ampeloprassum Linn.

Family : Liliaceae.

Distribution : E. Himalaya in Arunanchal Pradesh from 500-1,500m.

Part Used : Bulbs.

Folk Use : Bulbs used as vegetable by the local tribals.

Reference : Arora (1985), Rana *et al.* (2005), Samant *et al.* (2001).

Allium ascalonicum Linn.

Family : Liliaceae.

English, Hindi and Regional Names:

Eng. : Eschallot, the Shellot;
Hindi : Ekakndalasum, Gandana;
Beng. : Gundhun.

Distribution : Widley cultivated all over India.

Part Used : Bulbs.

Folk Uses : Bulbs used for pickling and for flavouring curries, as the flavour is similar to that of garlic.

Active Constituents : Chief constituents responsible for the flavour and fragrance of oil extracted from bulbs are dimethyl disulphide, 3,4-dimethyl thiophene ,dimethyl trisulphide, dipropyl disulphide, dipropyl trisulphide, methyl-cis-propenyldisulphide, methyl propyl trisulphide, 1-methylthiopropyl ethyl disulphide, 2- n-hexyl-5-methyl-2,3-dihydrofuran-3-one,propyl- cis-propenyl disulphide, propyl-trans-propenyl disulphide, 2-n-octyl-5-methyl-2,3-dihydrofuran-3-one and quercitol, its 3,4'-diglucoside and 7,4'-diglucoside (I).

References : Ambasta (1986), Anonymous (1948-1976), Arora (1985), Kirtikar & Basu (1984), Mehtani & Ingle (1999), Rana *et al.* (2005), Rastogi & Mehrotra (1990), Watt (1889-1896).

Allium blandum Wall.

Family : Liliaceae.

Distribution : Trans-N.W. Himalaya in Kashmir.

Reproductive Cycle : June -August.

Part Used : Onion - like bulbs.

Folk Uses : Egg-shaped, onion-like bulbs eaten and considered as very hot.

References : Badhwar & Fernandez (1964), Chandra (1997).

Allium carolinianum DC.

Syn : *Allium blandum* Wall.; *Allium thomsonii* Baker.

Family : Liliaceae.

Distribution : W. Himalaya in Kumaon & Garhwal: 3,300-4,000 m.

Parts Used : Rootstocks.

Folk Uses : Dried rootstocks used as spice and for flavouring soups and seasoning foods.

References : Bennet (1987), Kapur *et al.* (1996), Rana *et al.* (2005), Pangtey *et al.* (1982), Samant *et al.* (2001), Sood *et al.* (2001).

Allium cepa Linn.

(Pl. 1A, C)

Family :Liliaceae.

English, Hindi, Sanskrit and Regional Names:

Eng. : The onion;
Hindi : Piyaj;
Sans. : Dirghpatra, Mahakanda, Palandu, Rajapriya;.
Beng. : Piyaj;
Guj. : Dhungari;
Mar. : Kanda;
Tam. : Niruli;
Tel. : Niruli;
Mal : Chuvannaulli.

Distribution : Cultivated throughout India.

Parts Used : Bulbs (underground). Bulbils (aerial).

Folk Uses : Bulbs used as very important vegetable ingredient of a whole range of the cooking items. The characteristic spicy flavour, aroma and taste of onion bulbs make it a very important flavouring ingredient; and its oil immensely used in the canning industry to flavour soups, meat, sauces, dressings, etc.

Active Constituents : Constituents of bulbs are carbohydrates, 11.6; fats, < 0.1; moisture 86.3; proteins 1.2; vit. A, 25 I.U.; vit. B , 40 I.U.; vit. C, 11mg. /100g; and 0.005% of the essential oils (responible for its acrid taste and unpleasant odour). Compounds contributing to the flavour profile include thiopropanal, S-oxide, propyl - and propenyl containing di- and trisulphides and 2,4 - and 3,4- dimethyl thiophenes.

References : Abraham (1981), Agarwal (1986, 2003), Ambasta (1986), Chevallier (1996), Harborne & Baxter (2001), Kirtikar & Basu (1984), Mehtani & Ingle (1999), Prasad & Abraham (1984), Rastogi & Mehrotra (1990, 1991, 1995), Sharma & Singh (2001).

Allium consanguineum Kunth.

Syn : *Allium stracheyi* Baker.

Family : Liliaceae.

Distribution : N.W. Himalaya. (J. & K. & H.P.): 2,500 – 4,000 m.

Part Used : Bulbs.

Folk Uses : Bulbs used as vegetable by the locals.

References : Arora & Pandey (1996), Bennet *et al.* (1991), Chandra (1997), Samant *et al.* (2001).

Allium govanicum Wall. ex Baker

Family: Liliaceae.

Distribution : Temperate Himalaya. Kumaon westwards to Kashmir.

Part Used : Rootstocks.

Folk Use : Dried rootstocks used as spices.

References : Bennet (1987), Chandra (1997), Koelz (1979), Negi & Pant (1992), Pangtey *et al.* (1982), Rana *et al.* (2005).

Allium humile Kunth.

Family : Liliaceae.

Regional Name : ***Lahul*** : Skotche.

Distribution : Temperate Himalaya (Kumaon westward to Kashmir): 3,200- 3,700 m.

Reproductive Cycle : May-June.

Part Used : Bulbs.

Folk Use : Globose bulbs garlic smelling and used as flavouring material.

References : Badhwar & Fernandez (1964), Rana *et al.* (2005), Samant *et al.* (2001), Singh *et al.* (2001a).

Allium jacquemontii Kunth.

Family : Liliaceae.

Regional Name :

Mandi : Dune.

Distribution : N.W. Himalaya (J. & K. and H.P.): 3,300-5,500 m.

Part Used : Bulbs.

Folk Uses : Bulbs and leaves used as spice and condiments by locals but less commonly used.

References : Samant *et al.* (2001), Singh (1996).

Allium pezqualskinum Regal.

Family : Liliaceae.

English Name :

Eng. : Wild Garlic.

Distribution : Trans-N.W. Himalaya (J. & K.) upto 3,500 m.

Part Used : Bulbs.

Folk Use : Young bulbs used as condiments for flavouring food items. Bulbs dried or made into paste for winter use.

Reference : Singh *et al.* (2001a).

Allium platyspathum Shrenk.

Family : Liliaceae.

Distribution : Trans-N.W. Himalaya (J. & K. to H.P.): 3,000-4,000m.

Part Used : Bulbs.

Folk Use : Bulbs and leaves used as vegetable by the local tribals.

Reference : Samant *et al.* (2001).

Allium porrum Linn.

Family : Liliaceae.

Regional Name :

Beng. : Paru.

Distribution : Cold climatic regions of the whole Himalaya.

Part Used : Bulbs.

Folk Uses : Green bulbs used as a spice and condiment for flavouring curries.

Active Constituents : Bulbs rich in carbohydrates, 17.2; fats, 0.1; minerals, (Ca., 0.05; P., 0.07; S., 0.06-0.072 %, Fe, 2.3mg/100); moisture, 78.9; proteins , 1.8%; vit. A 30 I.U.; vit.B, 75 I.U.; and vit. C 11mg/ 100mg.

References : Agarwal (2003), Anonymous (1948-1976), Arora (1985).

Allium pratii C.H.Wright.

Syn : *Allium victoralis* Hook. var. *angustifolium* ; *A.victoralis* non Linn. Hook.

Family : Liliaceae.

Distribution : C. Himalaya in Sikkim.

Reproductive Cycle : April-August.

Part Used : Bulbs.

Folk Use : Bulbs used for seasoning curries in Sikkim.

References : Badhwar & Fernandez (1964), Bennet (1987), Chandra (1997), Rana *et al.* (2005).

Allium rubellum M. Bieb.

Family : Liliaceae.

Hindi Names :

Hindi : Barni Pyaj, Chiri-pyaji, Jangli Pyaj.

Distribution : Common in N.W. Himalaya (Kashmir to Kumaon): 1,500 –3,500m.

Reproductive Cycle : May- September.

Part Used : Bulbs.

Folk Uses : Bulbs eaten raw or cooked as vegetable.

References : Arora & Pandey (1996), Chandra (1997), Negi & Pant (1992), Rana *et al.* (2005), Samant *et al.* (2001), Watt (1889-1896).

Allium sativum Linn.

(Pl. 1B, D)

Family : Liliaceae.

English, Hindi and Sanskrit Names:

Eng. : Churl's Treacle, Garlic, Poor Man's Treacle;
Hindi : Lahsan, Lasan;
Sans. : Arishtha, Bhutabhna, Grinjana, Lashuna, Rasona, Vatari.

Distribution : Widely cultivated all over India.

Parts Used : Bulbs or Cloves (underground). Bulbils (aerial).

Folk Uses : Bulbs and bulbils used as spices and condiments for seasoning and garnishing. Its oil particularly relished in meat preparations, soups, canned products, sauces, salad dressing and for flavouring vinegar.

Active Constituents : Bulbs rich in carbohydrates, 29.0; fats, 0.1; minerals, (Ca and P); moisture, 62.2; proteins, 6.3; vitamins A, B, C and E.

Important flavouring components of fresh garlic oil are allicin (2-propenyl propane thiosulphate), allin, allinase, scordinins, and selenium.

References : Abraham (1981), Agarwal (1986), Ambasta (1986), Bhattacharjee (1998), Chevallier (1996), Harborne & Baxter (2001), Kirtikar & Basu (1984), Mehtani & Ingle (1999), Rastogi & Mehrotra (1991,1993,1995), Sharma & Singh (2001), Singh & Pandey (1998), Singh *et al.* (1996), Watt (1889-1896).

Allium schoenoprassum Linn.

Family : Liliaceae.

English Name :

Eng. : Cives, The Chive.

Distribution : N.W. and W. Himalaya (Kashmir to Kumaon): 2,400-3,300 m.

Part Used : Bulbs.

Folk Uses : Bulbs used for seasoning and flavouring foods and also used as salad.

References : Ambasta (1996), Anonymous (1948-1976), Arora & Pandey (1996), Chandra (1997), Kirtikar & Basu (1984), Rana *et al.* (2005), Rastogi & Mehrotra (1991, 1993), Watt (1889-1896).

Allium semonovii Regal.

Family : Liliaceae.

English Name :

Eng. : Wild Garlic.

Distribution : Trans-N.W. Himalaya (J. & K. and H.P.).

Part Used : Bulbs.

Folk Uses : Bulbs edible.

References : Singh *et al.* (2001a), Rana *et al.* (2005), Samant *et al.* (2001).

Allium stracheyi Linn.

Family : Liliaceae.

Distribution : N.W. Himalaya (Kumaon, Garhwal region): 3,500-4,500m.

Part Used : Bulbs.

Folk Uses : Bulbs edible.

References : Arora & Pandey (1996), Samant *et al.* (2001).

Allium tuberosum Roxb.

Family : Liliaceae.

Regional Names :

Ass. : Aphioa-naharu;

Megh. : Jyllang.

Distribution : Assam, Khasia Hills. N.E. Himalaya; semicultivated in Bengal.

Part Used : Roots. Bulbs.

Folk Uses : Roots used as vegetable and salad. Bulbs highly aromatic and edible.

References : Agarwal (2003), Arora (1985), Borthakur (1996), Joseph & Kharkongor (1996), Rana *et al.* (2005).

Allium victoralis Linn.

Family : Liliaceae.

Distribution : Temperate Himalaya (Kashmir eastwards to Sikkim): 2,100- 3,900m.

Part Used : Bulbs.

Folk Uses : Bulbs used for flavouring the vegetables, curries and food items.

References : Arora & Pandey (1996), Chandra (1997), Samant *et al.* (2001).

Allium wallichii Kunth.

Family : Lliliaceae

Distribution : Temperate Himalaya and N.W. Himalaya from J. & K. to Nepal: 2,500-4,500m.

Reproductive Cycle : July-August.

Part Used : Bulbs.

Folk Use : Bulbs pounded with spices to increase the aroma and flavour of the food.

References : Bennet *et al.* (1991), Chandra (1997), Rana *et al.* (2005), Samant *et al.* (2001), Singh *et al.* (1996).

Allophylus serratus Radlk.

Family : Sapindaceae.

Hindi Name :

Hindi **:** Tipani.

Distribution : Subtropical regions throughout India.

Part Used : Roots.

Folk Use : Roots used as famine food.

Reference : Agarwal (2003).

Alocasia cucullata Schott.

Family : Araceae.

Regional Name :

Mani. **:** Shingju Paan.

Distribution : N.E. Himalaya (Manipur and Assam).

Part Used : Tuberous Corms.

Folk Uses : Tubers used as vegetable after boiling and yield 'Taro' and 'Yam' for N. E. tribes. Corm of the plant used to prepare the traditional salad 'Shingju' and 'Iromba' and also cooked in fish curries.

References : Ambasta (1986), Anonymous (1948-1976), Arora (1981, 1985), Singh *et al.* (2001b).

Alocasia indica (Roxb.) Schott.

Syn : *Alocasia macrorrrhiza* (L.) G.Don.

Family : Araceae.

English, Hindi, Sanskrit and Regional Names:

Eng. : Taro, Yam;
Hindi : Mankanda;
Sans. : Manaka;
Beng. : Man-kachu.

Distribution : N.E. Himalaya. Bengal.

Part Used : Rhizomes.

Folk Uses : Rhizomatous portion used as vegetable after boiling. Also it is pulped for the extraction of nutritious starch and yield 'Taro' and 'Yam' for the N.E. tribals.

References : Agarwal (2003), Ambasta (1986), Arora (1981), Bennet *et al.* (1991).

Alocasia macrorrhiza Schott.

Syn : *Alocasia indica* (Lour) Spach.

Family : Araceae.

English, Hindi, Sanskrit and Regional Names:

Eng. : Giant Taro;
Hindi : Bara mankachu, Mankanda;
Sans. : Manaka;
Ass. : Baro-mankachu;
Beng. : Mankachu;
Kan. : Mannaka;
Mar. : Abu;
M.P. : Kola-kachu;
Mal. : Marambu;
Tam. : Merukonkelangu.

Distribution : C. India. N.E. Himalaya: 1,000-2,000m.

Reproductive Cycle : July-September.

Part Used : Tuberous Rhizome.

Folk Uses : Tuberous underground rhizome used as vegetable after boiling to remove the crystals of the calcium oxalate and hydrocyanic acid. Tubers may also be preserved for months after drying and eaten in curries, 'baddies', etc.

Active Constituents : Rhizome contains about 20.45% of starch.

References : Ambasta (1986), Arora & Pandey (1996), Borthakur (1996), Samant *et al.* (2001).

Remarks : Crystals of calcium oxalate and hydrocyanic acid are poisonous and responsible for acrid taste. These should be removed before eating.

Alpinia blepharocalayx Schum.

Family : Zingiberaceae

Regional Name :

Ass. : Tara.

Distribution : N.E. Himalaya.

Part Used : Rhizomes.

Folk Use : Rhizomes used as mouth freshner.

Reference : Tripathi & Goyal (2001).

Alpinia conchigera Griff..

Family : Zingiberaceae.

Part Used : Rhizomes.

Folk Use : Rhizomes used as condiment.

References : Ambasta (1986), Arora (1981), Arora & Pandey (1996), Bhattacharjee (1998).

Alpinia galanga (Linn.) Willd.

Syn : *A. rheedii* Wight; *Amomum galanga* Lour.

Family : Zingiberaceae.

English, Hindi, Sanskrit and Regional Names:

Eng. : Java Galangal, The Greater Galangal;

Hindi : Kulanjan;

Sans. : Aruna, Dhumala, Elaparni, Gandhamula, Kulanjan;

Ass. : Kanghu, Korphul;

Guj. : Kulinjan;

Kan. : Dumbarasme;

Mal.	:	Kolangbi, Pertta;
Man.	:	Pullei;
Mar.	:	Kosht kulinjan;
Tam.	:	Pura tallai.

Distribution : Often cultivated in N.E. Himalaya and S. W. India.

Reproductive Cycle : April-May.

Part Used : Rhizomes.

Folk Uses : Rhizomes used as a substitute of ginger for seasoning fish, meat and food. Also, it is used for pickling and in the preparation of traditional food 'Iromba' alongwith the other vegetables by the N.E. tribals.

Active Constituents : Rhizomes rich in starch and essential oils. Its active principles are due to alpinin, galangin, kaemferide, hydroxy-cinnamaldehyde, cineol and pungent oily substance galangol.

References : Agarwal (2003), Ambasta (1986), Arora (1981), Arora & Pandey (1996), Bennet *et al.* (1991), Bhattacharjee (1998), Borthakur (1996), Islam (1996), Jain (1995), Kirtikar & Basu (1984), Mittre (1981), Raju (2000), Rastogi & Mehrotra (1993, 1995), Saklani & Jain (1994), Singh *et al.* (2001b), Sinha (1996), Tripathi & Goyal (2001), Watt (1889-1896).

Alpinia nigra (Gaertn.) Burtt.

Family : Zingiberaceae.

Regional Names :

Megh.	:	Gong (Garo tribe);
Mani.	:	Pullei.

Distribution : E. Himalaya, particularly in Meghalaya.

Part Used : Rhizomes.

Folk Uses : Rhizomes edible among the 'Garo' tribes of 'Meghalaya' in E. India, and used in the preparation of 'Iromba' which is a special delicacy in major meals prepared from the plant ingredients boiled alongwith the fruits of the *Capsicum annum* Linn. (Chilli). These are made into a paste with a little roasting and then mixed with 'Nagari' ('Nagri' refer to the fermented preparation of the following three species of the fish namely *Punitius sophore* Ham.; *P.chola* Ham and *P.ticto* Ham.). Use of this fermented product enhances the palatability of the dish. 'Iromba' may also be prepared without 'Nagari'.

References : Jain (1995), Rao & Shanpru (1981), Singh et al. (2001b).

Alpinia speciosa K. Schum.

Family : Zingiberaceae.

English, Hindi and Regional Names:

Eng. : Light Galangal;
Hindi : Ligandh-tora;
Beng. : Punag Champa.

Distribution : N.E Himalaya. Also, cultivated in the gardens of India.

Parts Used : Rhizomes. Bulbs.

Folk Uses : Rhizomes eaten boiled or roasted or as vegetable in E. India. It is also used as a substitute for ginger and used as a spice and condiment.

References : Agarwal (2003), Anonymous (1948-1976), Arora (1981), Borthakur (1996), Jain (1995), Kirtikar & Basu (1984), Mittre (1981), Watt (1889-1896).

Alpinia zerumbet (Pers.) Burtt & R. M. Smith

Syn : *Alpinia speciosa* K. Schum.; *Alpinia nutans* Rosc.

Family : Zingiberaceae.

English and Regional Names:

Eng. : The light galangal;
Beng. : Punag Champa;
Tam. : Sit-ruthi.

Distribution : N.W. Himalaya.

Reproductive Cycle : May-August.

Part Used : Rhizomes.

Folk Uses : Rhizomes used as vegetable, condiments and substitute of *Alpinia galanga* for seasoning fish, curries and also for pickling.

References : Ambasta (1986), Arora (1981), Bennet *et al.* (1991), Tripathi & Goyal (2001).

Althaea officinalis Linn.

Family : Malvaceae.

English Name :

Eng. : Marsh Mallow.

Distribution : N. W. Himalaya (Kashmir, Panjab, H.P.).

Part Used : Roots.

Folk Uses : Roots used as salad and for brewing tea.

Active Constituents : Roots contains 37% starch.

References : Ambasta (1986), Anonymous (1948-1976).

Amischophacelus axillaris (L.) Rola Rao & Kamm.

Syn : *Cyanotis axillaris* (L.) R.&S.

Family : Commelinaceae.

Hindi and Regional Names:

Hindi : Baghanulla, Soltraj;
Mar. : Nirphulli;
Tam. : Nirpolli, Vallukkeiphul;
Tel. : Golagundi.

Distribution : Throughout India.

Reproductive Cycle : October-December.

Part Used : Roots.

Folk Use : Roots alongwith its leaves and seeds eaten at the time of food scarcity.

References : Chandra (1997), Lancaster (1964).

Amomum janii Sunil & Ved Prakash

Family : Zingiberaceae.

Distribution : Almost throughout India.

Part Used : Rhizomes.

Folk Uses : Rhizomes used as spice and condiment.

References : Anonymous (1948-1976), Tripathi & Goyal (2001).

Amorphophalus bulbifer Bl.

Family : Araceae.

Distribution : N.E.India.

Reproductive Cycle : May-July.

Part Used : Corms.

Folk Use : Corms edible and cooked as vegetable.

Reference : Bennet *et al.* (1991).

Amorphophallus campanulatus (Roxb.) Blume *ex* Decne. (Pl. 2A)

Syn : *Amorphophallus paeoniifolius* (Dennst.) Nicolson.

Family : Araceae.

English, Hindi, Sanskrit and Regional Names:

Eng. : Elephant foot Yam;
Hindi : Arsaghana, Balukand, Zaminkand;
Sans. : Arsaghana;
H.P. : Pahari Suran;
Kan. : Suvarna gerdde;
Mal : Chenakarona- Kizhanna;
Mar. : Jangli Suran;
Tam. : Kurruna kalungu;
Tel. : Kunda- muncha -kunda.

Distribution : Humid parts of E. and W. India, Maharashtra and Goa. Also cultivated in some parts of India.

Reproductive Cycle : April -June.

Part Used : Corms.

Folk Uses : Corms after proper washing and boiling used as a starchy vegetable and cooked with tamarind or lemon or other acidulous spices to dispel any irritating substances present in raw state. The corms can be kept without rotting for five to six months under the ordinary conditions of room temperature.

Active Constituents : Corms comprise of carbohydrates, 18.4; fats, 0.1; moisture, 78.7; minerals (Ca, 50.00; P, 34.00, Fe ,0.6mg/100 mg) and proteins, 1.2 %; The yield of the starch is less than 4.4 %.

References : Ambasta (1986), Anonymous (1948-1976), Arora (1985), Arora & Pandey (1996), Bennet *et al.* (1991), Borthakur (1996), Rao & Henery (1996), Islam (1996), Rana (2004), Rana *et al.* (2005), Vartak (1981).

Amorphophallus dubius Bl.

Family : Araceae.

English, Sanskrit and Regional Names:

Eng. : Purple-stalked Dragon;
Sans. : Kanona kanda;
Mal. : Kattuchena;
Tam. : Kattukarnei.

Distribution : S. India. Kerala.

Reproductive Cycle : April - June.

Part Used : Corms.

Folk Use : Corms boiled in tamarind and used as vegetable.

References : Bennet *et al.* (1991), Rao (1914).

Amorphophallus rivieri Dur.

Family : Araceae.

Distribution : Semicultivated in the plains of India.

Part Used : Tuberous Corms.
Folk Use : Tuberous corms boiled, cooked and eaten.
Reference : Ambasta (1986).

Amorphophallus sylvaticus (Roxb.) Kunth.

Syn : *Synantheria sylvatica* (Roxb.) Schott.
Family : Araceae.

Hindi, Sanskrit and Regional Names:

Hindi : Vajra-kand;
Sans. : Vajra-kand;
Mar. : Vajra-mula.

Distribution : Nilgiri Hills. Madras (coastal area). Karnataka. Deccan Peninsula.
Part Used : Tubers.
Folk Use : Tubers edible after proper boiling and cooking.
References : Agarwal (1986, 2003), Bennet *et al.* (1991), Watt (1889-1896).

Ampelocissus araneosa Planch.

Syn : *Vitis araneosis* Dalz. & Gibs *ex* M.Law.
Family : Vitaceae.

Hindi Name :
Hindi : Kauraj.

Part Used : Tuberous Roots.
Folk Uses : Tuberous roots sliced and marketed under the name of 'Chamar - musli' and eaten raw or cooked.
Reference : Ambasta (1986).

Angelica archangelica Linn. var. *himalaica* (C.B. Clarke) Krishna & Badhwar

Syn : *Archangelica officinalis* Hoffm. var. *himalaica* C.B. Clarke.
Family : Apiaceae.

Distribution : Temperate Himalaya (Kashmir and Sikkim).

Part Used : Roots.

Folk Uses : Aromatic roots used as condiments for flavouring foods, confectionaries and as seasoning agent for curries, baked goods, gelatins and puddings. Also it finds application for flavouring alcoholic and nonalcoholic beverages.

Active Constituents : Roots mainly contain archangelenone, and its musk-like odour has been attributed to pentadecanolide.

References : Agarwal (2003), Ambasta (1986), Anonymous (1948-1976), Chandra (1997), Chevallier (1996), Harborne & Baxter (2001), Mehtani & Ingle (1999), Negi & Gaur (1994), Rastogi & Mehrotra (1990).

Angelica glauca Edgew. (Pl. 2B)

Family: Apiaceae.

English and Regional Names:

Eng. : Angelica;
H.P. : Chora;
Kash. : Chohore,Chura;
Pub. : Chora, Chura;
U.P. : Chorak; Gandhrayan.

Distribution : W. Himalaya (Kashmir to Shimla): 2,700-3,500 m.

Reproductive Cycle : July -August.

Part Used : Roots. Rhizomes.

Folk Uses : Roots added to food for flavour like that of celery in W. Himalaya. The essential oil extracted from its roots is used as a substitute of 'European Angelica' oil for use in confectionery and liquors. Dried rhizomes pounded with spices are mixed to different food preparations.

Active Constituents : The major flavouring agents of roots are angelicin, bergapten, xanthotoxin, angelicoside, archangelin and oxypeucedanin.

References : Agarwal (2003), Ambasta (1986), Anonymous (1948-1976), Arora & Pandey (1996), Bhattacharjee (1998), Bennet *et al.* (1991), Chandra (1997), Kapur *et al.* (1996), Mehtani & Ingle (1999), Mittre (1981), Rana *et al.* (2005), Rastogi & Mehrotra (1990), Samant *et al.* (2001), Sharma & Sood (1997), Singh (1996), Singh *et al.* (1996), Watt (1889-1896).

Angiopteris evecta (Forsk.) Hoffm.

Family : Marattiaceae.

Regional Name :

Ass. : Chainau – aukimu.

Distribution : N.E. Himalaya upto 2000 m.

Parts Used : Rhizomes.

Folk Uses : Underground rhizomes and the base of rachis consumed after thorough washing and boiling. Rhizomes also grounded into flour and eaten.

References : Borthakur (1996), Singh & Viswanathan (1996).

Anredera baselloides H.B. & K.

Syn : *Boussingaultia baselloides* (H.B.&K.).

Family : Basellaceae.

English Names :

Eng. : Lambs tail, Maderia vine, Mignonette vine.

Part Used : Roots.

Folk Uses : Roots used to flavour the local drinks and wines.

Reference : Ambasta (1986).

Antigonon leptopus Hook &Arn.

Family : Polygonaceae.

English Name :

Eng. : Coral Creeper.

Part Used : Tubers.

Folk Use : Tubers eaten.
Reference : Ambasta (1986).

Aponogeton crispum Thunb.

Family : Aponogetonaceae.

Distribution : Tropical Peninsular India.
Parts Used : Tuberous Rootstocks. Rhizomes.
Folk Uses : Tuberous rootstocks and rhizomes consumed as vegetable.
References : Ambasta (1986), Arora & Pandey (1996), Islam (1996).

Aponogeton natans (Linn.) Engl. & Krause

Syn : *Aponogeton monostachyon* Linn.
Family : Aponogetonaceae.

Hindi, Sanskrit and Regional Names:

Hindi : Ghechu;
Sans. : Kangii;
Mal. : Parraki zhangu;
Tam. : Kotti Kizhangu, Kotti kallangu;
Tel. : Kotti-gadda, Nama.

Distribution : Tropical Peninsular India.
Reproductive Cycle : Throughout the year.
Parts Used : Rootstocks.
Folk Uses : Starchy rootstocks considered good as potato and esteemed as great delicacy and usually eaten after boiling or cooking as vegetable.
References : Ambasta (1986), Anonymous (1948-1976), Arora & Pandey (1996), Bennet *et al.* (1991), Shortt (1887-1888), Subramanyam (1964).

Aponogeton undulatus Roxb.

Syn : *A. crispus* sensu Hook.f.; *A. microphyllum* Roxb.
Family : Aponogetonaceae.

Distribution : N.E. Himalaya. Throughout the greater parts of India.

Reproductive Cycle : June – October.

Part Used : Tuberous Rootstocks.

Folk Use : Tuberous rootstocks eaten.

References : Bennet *et al.* (1991), Subramanayam (1962).

Arisaema concinnum Schott. (Pl. 2C)

Family : Araceae.

Distribution : Temperate Himalaya from Panjab to Arunanchal Pradesh: 1,800-3,300m.

Reproductive Cycle : March – July.

Part Used : Tuberous Corms.

Folk Uses : Corms used as vegetable only after repeated boiling to remove the crystals of calcium oxalate. Also consumed as food during scarcity.

Active Constituents : Tuber comprises of albuminoids, 7.7; carbohydrates, 65.9; fats, 7.7; fibers 8.9; minerals in ash, 7.7 % (Ca., 1.2%; P_2O_5, 0.2%); and moisture, 8.5%.

References : Ambasta (1986), Anonymous (1948-1976), Arora & Pandey (1996), Chandra (1997).

Arisaema costatum (Wall.) Mart. *ex* Schott.

Family : Araceae.

Distribution : Sikkim Himalaya : 1,800- 2,000m.

Part Used : Tubers.

Folk Use : Tubers consumed.

Reference : Chandra (1997).

Arisaema curvatum (Hook.) Kunth.

Family : Araceae.

Distribution : Temperate Himalaya.

Part Used : Tuberous Roots.

Folk Use : Tuberous roots edible.

Reference : Watt (1889-1896).

Arisaema jacquemontii Bl.

Family : Araceae.

Distribution : Temperate Himalaya from Kashmir to Bhutan: 2,000-3,000 m.

Part Used : Tuberous Roots.

Folk Uses : Tuberous roots cooked, roasted and mixed with salt and pepper (chillies) and form an excellent preparation.

References : Arora & Pandey (1996), Chandra (1997), Pangtey *et al.* (1982), Rana *et al.* (2005), Samant *et al.* (2001).

Arisaema murrayi Hook.

Family : Araceae.

Hindi and Regional Names:

Hindi : Diva Baddad;
Bomb. : Dhudhda, Diwa, Kanala Sopha Kanda;
Mar. : Badad.

Distribution : W. Peninsula. Maharashtra.

Reproductive Cycle : June - October.

Part Used : Tuberous Corms.

Folk Use : Corms eaten as vegetable.

References : Chandra (1997), Datar & Vartak (1975), Gammie (1902), Vartak (1981).

Arisaema propinquum Schott.

Family : Araceae.

Distribution : N.W. Himalaya.

Part Used : Tuberous Corms.

Folk Use : Tuberous corms eaten as vegetable.

Reference : Arora & Pandey (1996).

Arisaema speciosum Mart.

Family : Araceae.

English, Hindi and Regional Names:

Eng. : Cobra Lily, Snake Lily;
Hindi : Kiralu ;
Pub. : Kiralu, Kiri ki Kuri, Samp ka Bhutta, Samp ki khumb.

Distribution : Temperate Himalaya (Kashmir to Kumaon) and Arunanchal Pradesh: 1,500- 3,000m.

Reproductive Cycle : May - July.

Part Used : Tuberous Corms.

Folk Uses : Corms fairly nutritious and consumed after repeated boiling to remove crystals of calcium oxalate.

Active Constituents : Tubers rich in albuminoids, 7.7; carbohydrates, 76.00; fats, 1.6; fibres, 6.1%; minerals in ash, 5.1 (Ca., 1.0%; P_2O_5, 0.5%) and moisture, 7.6%.

References : Agarwal (2003), Ambasta (1986), Arora & Pandey (1996), Chandra (1997), Rana *et al.* (2005).

Arisaema tortuosum Schott.

Family : Araceae.

English, Hindi and Regional Names:

Eng. : The Cobra flower;
Hindi : Gurien, Kiri ki Kuri, Samp ki khumb ;
Beng. : Birman;
Kan. : Awu -Mari -Gidda;
Mar. : Baddha, Dhudha, Sapkanda;
Santh. : Tuya -Jondra.

Distribution : Almost throughout subtropical and temperate Himalaya upto 2,800 m.

Reproductive Cycle : August-September.

Part Used : Stem Tubers.

Folk Uses : Tubers cooked as vegetable only after proper boiling and washing by the tribals. Also, used as famine food.

References : Agarwal (2003), Bennet *et al.* (1991), Chandra (1997), Gammie (1902), Krishna *et al.* (1999), Samant *et al.* (2001).

Remarks : Tubers acrid and poisonous and should only be consumed after proper boiling.

Arisaema tortuosum var. curvatum (Roxb.) Engl.

Syn : *Arisaema curvatum* (Roxb.) Kunth. non Hook. f.

Family : Araceae.

Regional Names :

Kash. : Surin;

Pub. : Dor, Kirakal, Kirki.

Distribution : C. and N.W.Himalaya. Punjab. Peninsular India.

Reproductive Cycle : July-August.

Part Used : Tubers.

Folk Uses : Tubers cooked as vegetable by the tribals. It is also used as famine food.

Reference : Chandra (1997).

Remarks : Tubers acrid and poisonous and consumed after proper boiling.

Armoracia rusticana Gaertn *et al.*

Syn : *A. lapathifolia* Gibb.; *Cochlea armoracea* Gartn *et al.*

Family : Brassicaceae.

English Name :

Eng. : Horse Raddish.

Part Used : Roots.

Folk Use : Roots used as condiments.

Reference : AmbastaM (1986).

Arnebia guttata Bunge.

Family : Boraginaceae.

Regional Names :

Lahul. : Demok.

Distribution : Trans-N.W. Himalaya (Lahul & Spiti Distt. of H.P.) upto 4,300 m.

Part Used : Roots.

Folk Uses : Red edible dye extracted from its roots used to colour the confectionaries and foods.

Reference : Singh *et al.* (2001a).

Arcticum lappa Linn.

Family : Asteraceae.

Regional Name :

Kash. *:* Cheron, Phaggarmul.

Distribution : N.W.Himalaya (Kashmir to Shimla): 1,800-2,700m.

Reproductive Cycle : July-September.

Part Used : Roots.

Folk Use : Roots peeled, boiled and eaten as vegetable.

Active Constituents : Analysis (only edible portion) of root reveals moisture, 73.7; proteins, 3.1%; starch and ash.

Reference : Chandra (1997).

Asparagus adscendens Roxb. (Pl. 3A)

Family : Liliaceae.

Regional Names :

Garh. *:* Jhirna;

Guj. *:* Ujlimusli.

Distribution : Tropical and subtropical parts of N.W. Himalaya upto 1,600 m.

Reproductive Cycle : March - June.

Part Used : Tuberous Roots.

Folk Uses : Tubers edible and used as vegetable. Roots pickled and used similar to 'Salap Misri' (*Orchis moschata* Linn.)

References : Agarwal (2003), Ambasta (1986), Anonymous (1948-1976), Bajpayee & Dixit (1996), Bennet *et al.* (1991), Islam (1996), Rana (2004), Rana *et al.* (2005).

Asparagus filicinus Buch.-Ham. *ex* Don.

Family : Liliaceae.

Regional Names :

H.P. *:* Alli-pali, Raunta, Sharanoi.

Distribution : Throughout Himalaya from Kashmir to Arunanchal Pradesh: 1,800-2,700 m.

Part Used : Tuberous Roots.

Folk Uses : Tuberous roots pickled and eaten in Kumaon.

Reference : Bennet *et al.* (1991).

Asparagus gonocladus Baker

Family : Liliaceae.

Hindi and Regional Names:

Hindi : Shakakul;
Beng. : Satmuli;
Guj. : Shatavari;
Mar. : Shatavarimul;
Tam. : Kilavari;
Tel. : Challa -gaddu, Pilipui –chara.

Distribution : Peninsular India from Konkan southwards in the Ghats: 1,200- 1,800 m. Plains of India.

Reproductive Cycle : July-September.

Part Used : Roots.

Folk Use : Roots edible and considered nutritious.

Reference : Chandra (1997).

Asparagus officinalis Linn.

Family : Liliaceae.

English, Hindi and Regional Names:

Eng. : Asparagus ;
Hindi : Nag- Down;
Beng. : Hillua, Hylyum.

Distribution : Often grown in Indian gardens.

Part Used : Roots.

Folk Use : Roots used as vegetable.

Active Constituents : Roots comprise of carbohydrates, 3.2; fats, 0.2;fibres, 0.7;minerals, 0.7 % (Ca, 1.0; P_2O_5, 0.039%; Fe, 0.96mg/100g, Cu, 0.14 mg/100g) moisture, 93.00; proteins, 2.2%; vit. A, 1400 I.U., vit. B_1, 180mg/100g, vit. B_2, 130 mg/100g and vit. C, 40mg/ 100g.

References : Agarwal (2003), Anonymous (1948-1976), Nayer *et al.* (1989-1994).

Asparagus racemosus Willd.

Family : Liliaceae.

Hindi, Sanskrit and Regional Names:

Hindi : Satmuli, Satwar;
Sans. : Satmuli, Satawari;
Kan. : Majjige - gadde;
Mar. : Aswal;
Mal. : Shatavali;
Tam. : Shimai- shadavari.

Distribution : C. India. Entire Himalaya. Tropical and subtropical regions throughout India.

Reproductive Cycle : July -November.

Part Used : Fasciculated Root Tubers.

Folk Uses : Root tubers eaten boiled, raw or in candid form. Also powdered roots taken with milk as tonic.

Active Constituents : Root tubers comprise of carbohydrates, 28.2g, fats, 0.3g; moisture, 68.5g; proteins, 1.2g/100g of edible portion.

References : Agarwal, (2003), Ambasta (1986), Anonymous (1948-1976), Arora & Pandey (1996), Bajpayee & Dixit (1996), Bennet *et al.* (1991), Das & Chattopadhyay (2001), Gupta (1962), Gupta & Kanodia (1968), Islam (1996), Krishna *et al.* (1999), Samant *et al.* (2001), Saxena (1979), Vartak (1981).

Asparagus sarmentosus Linn.

Family : Liliaceae.

Distribution : Grows throughout India.

Part Used : Fleshy Roots.

Folk Uses : Fleshy roots edible and used as famine food.

References : Agarwal (1986), Ambasta (1986), Anonymous (1948-1976), Arora (1985), Watt (1889-1896).

Asphodelus albus Boiss.

Family : Liliaceae.

Part Used : Bulbous Tubers.
Folk Use : Tubers eaten as famine food.
Reference : Watt (1889-1896).

Asphodelus fistulasus Linn.

Family : Liliaceae.

Distribution : E. Himalaya.
Part Used : Bulbs.
Folk Use : Bulbs eaten at the time of scarcity.
References : Anonymous (1948-1976), Arora & Pandey (1996).

Asphodelus tenuifolius Cav.

Family : Liliaceae.

Distribution : Garhwal Himalaya.
Part Used : Bulbs.
Folk Use : Bulbs eaten at the time of scarcity.
References : Arora & Pandey (1996), Gupta (1962).

Aspidopterys indica Willd.

Family : Malpighiaceae.

Distribution : N.E. Himalaya. Khasia hills.
Part Used : Root Tubers.
Folk Use : Root tubers eaten as food by the tribals.
Reference : Arora & Pandey (1996).

Beta vulgaris Linn.

Family : Chenopodiaceae.

English and Hindi Names:

Eng. : Beet root or Garden beet;
Hindi : Chukander.

Distribution : Widely cultivated throughout India.
Reproductive Cycle : October – April.
Part Used : Roots.
Folk Uses : Roots eaten as salad and vegetable. At many places, roots employed industrially for manufacturing beet sugar.
Active Constituents : Various constituents of roots are ash, 8.5; fibres, 1.1; moisture, 76.6; oils, 0.1; proteins 1.1; soluble carbohydrates, 20.4; minerals, (K_2O, 40.00%, P_2O_5, 10-26 % Fe, 1.0mg/100g), vit. A, 21000 I.U.; vit. B_1, 70 I.U. and vit. C, 88mg /100g.
References : Agarwal (2003), Ambasta (1986), Anonymous (1948-1976), Rana (2004).

Blechnum orientale Linn.

Family : Polypodiaceae.

Hindi and Regional Names:

Hindi : Hathathjori;
Mizo. : Vomban (Mizo).

Distribution : N.E Himalaya upto 4,000 m.
Part Used : Rhizomes.
Folk Use : Rhizomes cooked and eaten as vegetable.
References : Ambasta (1986), Anonymous (1948-1976), Singh & Viswanathan (1996), Sinha (1996).

Borassus flabellifer Linn.

Syn : *Borassus flabelliformis* Murr.
Family : Aracaceae.

English, Hindi and Regional Names:

Eng. : Palmarya Palm;
Hindi : Tar;
Bomb. : Tad, Tad-mar;
Tam. : Pannei;
Tel. : Karatalamu, Tatichettu.

Distribution : Coastal areas of Peninsular India. Rajasthan, Madras, Bihar and Bengal (dry parts).

Part Used	:	Fusiform Roots.
Folk Use	:	Fleshy roots eaten by the tribals.
References	:	Agarwal (2003), Arora & Pandey (1996), Gammie (1902), Gupta & Kanodia (1968), Shortt (1887-1888), Watt (1889-1896).

Boussingaultia baselloides H.B. &K.

Family : Basellaceae.

Distribution	:	S. India.
Part Used	:	Tubers.
Folk Uses	:	Tubers considered rich in proteins and eaten as vegetable.
Active Constituents	:	Fresh tubers contain ash, 1.35; crude fibres, 0.4; fats, 0.27; moisture, 85.00; nitrogen free extract, 10.50; and nitrogenous substances, 2.3%.
References	:	Agarwal (2003), Anonymous (1948-1976).

Brachystelma laevigatum Hook. f.

Family : Asclepiadaceae.

Part Used	:	Tubers.
Folk Uses	:	Tubers eaten after removal of the resinous layer.
Reference	:	Ambasta (1986).

Brasenia shreberi J.F.Geml.

Syn : *Brasenisa peltata* Puresh.

Family : Cabombaceae.

Distribution	:	Meghalaya. Khasia Hills: 1,000 - 1,300 m.
Part Used	:	Tuberous Rootstocks.
Folk Use	:	Tuberous rootstocks used as food.
Reference	:	Behnet *et al.* (1991).

Brassica campestris sub sp. *napus* Linn.

Family : Brassicaceae.

Distribution	:	Cultivated all over India.
Part Used	:	Roots.

Folk Uses : Roots edible.

References : Agarwal (2003), Ambasta (1986), Anonymmous (1948-1976).

Brassica campestris Linn. var. *dichotoma* Watt.

Family : Brassicaceae.

English and Hindi Names:

Eng. : Mustard;

Hindi : Kali Sarson.

Distribution : Widely cultivated throughout N.W. India.

Part Used : Roots.

Folk Uses : Roots dried and used as vegetable.

References : Agarwal (2003), Ambasta (1986), Anonymous (1948-1976).

Brassica campestris Linn. var. *sarson* Prain

Family : Brassicaceae.

English and Hindi Names:

Eng. : Mustard;

Hindi : Sarson.

Distribution : Cultivated throughout India.

Part Used : Tuberous Roots.

Folk Uses : Roots dried and used as vegetable.

References : Agarwal (2003), Ambasta (1986), Anonymous (1948-1976).

Brassica oleracea Linn. convar. *botrytis*

Family : Brassicaceae.

English and Hindi Names:

Eng. : Brocooli, Cauliflower ;

Hindi : Gobhi, Phoolgobhi.

Distribution : Grown throughout India

Part Used : Roots.

Folk Uses : Roots edible.

References : Ambasta (1986), Agarwal (2003).

Brassica oleracea Linn. var. *capitata*

Family : Brassicaceae.

English and Hindi Names:

Eng. : Cabbage;
Hindi : Bandhgobhi.

Distribution : Grown throughout India.
Part Used : Roots.
Folk Uses : Roots edible.
Reference : Agarwal (2003).

Brassica rapa Linn. (Pl. 3B)

Family : Brassicaceae.

English and Hindi Names:

Eng. : Turnip ;
Hindi : Salgam, Shalgam.

Distribution : Cultivated throughout India.
Reproductive Cycle : September- April (Next year).
Part Used : Roots.
Folk Uses : Globose roots used as vegetable and salad throughout India, and also pickled.
Active Constituents : Fresh turnip root contains carbohydrates, 7.6; fats 0.2; minerals, 0.6% (Ca, 40.00%, P_2O_5, 10.00-26.01%. Fe, 1.0mg/100g); moisture, 91.0; proteins 0.5%; vit A, 21000 I.U.; vit. B1, 70 I.U. and vit. C, 88mg / 100g.
References : Agarwal (2003), Ambasta (1986), Anonymous (1948-1976).

Brassica tournefortii Gouan.

Family : Brassicaceae.

English and Hindi Names:

Eng. : Panjab Rai;
Hindi : Jangli Shalgam.

Distribution : C. and W.Himalaya.
Part Used : Roots.

Folk Uses : Roots used as vegetable and also for extracting medicinal oil.

References : Agarwal (2003), Anonymous (1948-1976).

Bunium persicum (Boiss.) Fedts.

Family : Apiaceae.

English and Hindi Names:

Eng. : Black Caraway ;
Hindi : Kalazira.

Distribution : N.W. Himalaya especially in Lahaul (H.P.).

Part Used : Tubers.

Folk Use : Starchy tubers eaten by the local tribes (Lahaulian and Spitians).

References : Arora and Pandey (1996), Rana *et al.* (2005).

Bupleurum candolli Wall. *ex* DC.

Family : Apiaceae.

Distribution : Tropical Himalaya.

Part Used : Roots.

Folk Uses : Roots sweet in taste and edible.

Reference : Arora & Pandey (1996).

Bupleurum falcatum Linn.

Family : Apiaceae.

English Name :

Eng. : Hare's ear.

Distribution : N.E. and N.W. Himalaya (Kashmir to Bhutan).

Part Used : Roots.

Folk Uses : Boiled roots eaten by Khasi's tribe.

References : Agarwal (2003), Ambasta (1986), Anonymous (1948-1976), Arora (1981), Arora & Pandey (1996).

Bupleurum falcatum Linn. var. marginatum Wall.

Family : Apiaceae.

Hindi Name :

Hindi : Kalizewar.

Distribution : N.E. and N.W. Himalaya from Kashmir to Khasia Hills.

Part Used : Roots.

Folk Uses : Roots edible.

Reference : Arora & Pandey (1996).

Butea monosperma (Lam.) Kuntze.

Syn :*Butea frondosa* Koeig. *ex* Roxb.

Family : Papilionaceae.

English, and Hindi Names:

Eng. : Flame of the forest;
Hindi : Dhak, Palash.

Distribution : Throughout India.

Part Used : Roots.

Folk Uses : Roots edible. Powdered dried roots mixed with cornflour for making bread.

Reference : Arora & Pandey (1996).

Calamus tenuis Roxb.

Family : Arecaceae.

Distribution : Subtropical Himalaya.

Part Used : Rhizomes.

Folk Use : Rhizomatous portion edible particularly in the N.E. region of India.

Reference : Arora & Pandey (1996).

Caralluma ascendens (Roxb.) Haw. var. *fimbricata*

Syn : *Caralluma fimbricata* Roxb.

Family : Asclepiadaceae.

English, Hindi and Regional Names :

Eng. : Black Caraway ;
Hindi : Kalazira;
Mar. : Makad Sheguli.

Distribution : W. India. Rajasthan. Deccan: 600-900m.
Part used : Roots.
Folk Use : Roots consumed after mixing with salt and pepper.
References : Gammie (1902), Watt (1889-1896).

Caralluma tuberculata Br.

Family : Asclepiadaceae.

Distribution : Trans-N.W.Himalaya (J. & K. & H.P.): 1,000-1,800 m.
Part Used : Roots.
Folk Use : Roots edible.
Reference : Samant *et al.* (2001).

Careya arborea Roxb.

Family : Lecythidaceae.

English, Hindi and Regional Names:

Eng. : Kumbi (Trade Name);
Hindi : Kumbhi;
Kan. : Doddala, Kaval;
Mar. : Alam-Pelu, Kumbia;
Tam. : Ayma;
Tel. : Arya.

Distribution : Peninsular India. Maharashtra. Goa.
Part Used : Root-bark.
Folk Uses : Root-bark edible.
References : Bennet *et al.* (1991), Vartak (1981).
Remarks : Considered poisonous and eaten only after proper boiling.

Carum atteunata Hook.

Family : Apiaceae.

Distribution : C. India. E. and W. Ghats.
Part Used : Tubers.
Folk Uses : Tubers edible.
Reference : Krishna *et al.* (1999).

Carum bulbocastanum W. Koch.

Syn : *Carum nigrum* Royle.

Family : Apiaceae.

English and Hindi Names:

Eng. : Black Caraway ;
Hindi : Kala -zira, Shah- zira.

Distribution : W. Himalaya from Kashmir to Garhwal.

Part Used : Starchy tubers.

Folk Uses : Starchy tubers used as vegetable and salad.

References : Agarwal (2003), Ambasta (1986), Anonymous (1948-1976).

Carum carvi Linn.

Syn : *Carum copticum* Linn.; *Carum nigrum* L.; *Apium carv* Crantz.; *Sesali carvi* Lam.

Family : Apiaceae.

English, Hindi, Sanskrit and Regional Names:

Eng. : Caraway;
Hindi : Shah -zira;
Sans. : Sushavi;
Beng. : Jira;
Kash. : Gunyun;
Pub. : Jira-siah;
Tam. : Shimai.

Distribution : Temperate regions of India.

Part Used : Roots.

Folk Uses : Roots eaten as vegetable.

References : Agarwal (2003), Ambasta (1986), Arora & Pandey (1996), Rana *et al.* (2005).

Carum pussila Wight & Arn.

Family : Apiaceae.

Distribution : Temperate regions of India.

Part Used : Tubers.

Folk Use : Tubers edible.

Reference : Ambasta (1986).

Ceropegia bulbosa Roxb.

Family : Asclepiadaceae.

Hindi and Regional Names:

Hindi	:	Khapparkaddu ;
Bomb.	:	Pallala -tombou.
Mar.	:	Khantodi, Khapparkaddu;
Pub.	:	Galot;
Tam.	:	Hallik.

Distribution : Gangetic Plains. Hilly tracts of C. India. W. Ghats.

Part Used : Root Tubers.

Folk Uses : Tubers eaten raw or boiled after mixing with either milk and sugar or with salt and spices.

Active Constituents : Root tuber analysis shows albuminoids, 3.4; ash, 9.43; carbohydrates, 66.00; fats, 3.3; fibres, 12.6; and moisture, 5.2%. Its bitterness is due to ceropegine.

References : Agarwal (2003), Ambasta (1986), Arora & Pandey (1996), Anonymous (1948-1976) Bennet *et al.* (1991), Gammie (1902), Gupta & Kanodia (1968), Rana *et al.* (2005), Saxena (1979), Vartak (1981), Watt (1889-1896).

Remarks : Roots made edible by boiling to remove its bitterness.

Ceropegia evansii Mac. Cann.

Family : Asclepiadaceae.

Distribution : Peninsular India. Maharashta: 600-900 m.

Reproductive Cycle : July-October.

Part Used : Root tubers.

Folk Use : Tubers edible.

References : Chandra (1997), Nayer *et al.* (1989-1994).

Ceropegia hirusta Wt. & Arn.

Family : Asclepiadaceae.

Hindi Name :

Hindi *:* Khantali.

Distribution : Humid tracts of E. and W. Ghats.
Part Used : Tubers.
Folk Uses : Tubers edible particularly in Rajasthan.
Reference : Arora & Pandey (1996).

Ceropegia lawii Wt. & Arn.

Family : Asclepiadaceae.

Distribution : Maharashtra. Goa.
Part Used : Root Tubers.
Folk Uses : Root tubers edible.
References : Arora & Pandey (1996), Vartak (1981).

Ceropegia pusilla Wt. & Arn.

Family : Asclepiadaceae.

Distribution : Nilgiri and Annamalai Hills.
Part Used : Tubers.
Folk Use : Tubers eaten at the time of scarcity.
References : Anonymous (1948-1976), Bennet *et al.* (1991).

Ceropegia tuberosa (Roxb.)Wall.

Family : Asclepiadaceae.

Hindi Name:

Hindi *:* Pataltumbi.

Distribution : S. India. Maharashtra. W. Rajasthan
Part Used : Tubers.
Folk Uses : Tubers edible and eaten raw or roasted; and also used as tonic.
References : Agarwal (2003), Ambasta (1986), Saxena (1979).

Chaerophyllum acuminatum DC.

Family : Apiaceae.

Regional Name :

Raj. : Khadula.

Distribution : Trans-N.W. Himalayan regions of J. & K.: 2,000-3,200 m.

Part Used : Roots.

Folk Use : Roots edible.

Reference : Samant *et al.* (2001).

Chaerophyllum reflexum Lindl.

Family : Apiaceae.

Distribution : Himalaya from H.P. to West Bengal: 1,500-2,700m.

Reproductive Cycle : June-July.

Part Used : Small Tubers.

Folk Uses : Aromatic small tubers eaten raw as well as cooked.

References : Chandra (1997), Pangtey *et al.* (1982).

Chaerophyllum arundinaceum Baker

Family : Liliaceae.

Hindi Name :

Hindi : Biskandri, Safed musli.

Distribution : Cultivated in gardens for its ornamental flowers especially in E. Himalaya. Bihar.

Part Used : Roots.

Folk Uses : Roots edible and used as tonic.

References : Agarwal (1986, 2003), Arora & Pandey (1996), Bennet *et al.* (1991).

Chlorophytum tuberosum Baker.

Family : Liliaceae.

Regional Name :

Bomb. : Kolu, Sevani, Safed mosali.

Distribution : C. India. Deccan Peninsula.

Reproductive Cycle : July-August.
Parts Used : Roots. Bulbs.
Folk Uses : Dried roots and bulbs pounded into flour alongwith its leaves for making bread.
References : Ambasta (1986), Arora & Pandey (1996). Bennet *et al.* (1991). Gammie (1902).

Cichorium intybus Linn.

Syn : *C. divaricatum* Heldr. *ex* Nym.
Family : Asteraceae.

English, Hindi and Regional Names:

Eng. : Chicory, Wild Endive, Succory ;
Hindi : Hinduba, Kasni Kashini;
Beng. : Kasni;
Guj. : Kasni;
Kash. : Lishkan;
Mal. : Mumchetti ki Zhangri;
Pub. : Gulhand;
Tam. : Kashmiri Keeral;
Tel. : Kasini -vittalu.

Distribution : Entire Himalaya. Most parts of India.
Reproductive Cycle : April-May.
Part Used : Dried Roots.
Folk Use : Dried roots roasted and powdered for mixing with coffee.
Active Constituents : Fresh roots yield water, 77.00; gummy matter, 7.5; glucose, 1.1; bitter extractive, 4.0; fats, < 0.6; cellulose, inulin fibres, 9.0 and ash, 0.8% rich in potash.
References : Anonymous (1948-1976), Chandra (1997).

Cinnamomum verum J.S. Presl.

Syn : *C. zeylanicum* Blume.
Family : Lauraceae.

English , Hindi and Sanskrit Names:

Eng. : Ceylon Cinnamon, True Cinnamon;
Hindi : Dalchini, Darchini;
Sans. : Bahugandha, Chola, Darushila, Hridya, Tamalapatra.

Distribution : S. India upto 1,500 m. Also cultivated in certain parts of India.
Part Used : Root-bark.
Folk Use : Root-bark employed as spices.
Active Constituents : These are á-pinene, á-phellandrene, p-cymene, linalool, cinnamaldehyde (30.2%) (bark oil).
References : Bhattacharjee (1998), Chevallier (1996), Greenish (1999), Jain (1968), Kirtikar & Basu (1984), Lalramnghinglova (2002), Mehtani & Ingle (1999), Rastogi & Mehrotra (1993,1995), Saklani & Jain (1994).

Cirsium argyracanthus DC.

Family :Asteraceae.

Distribution : N.W. Himalaya (Kumaon and Garhwal): 1,000-2,000 m.
Part Used : Roots.
Folk Use : Roots used as vegetable by the locals.
Reference : Samant *et al.* (2001).

Cirsium wallichii DC.

Family : Asteraceae.

Distribution : Tropical Himalaya.
Part Used : Tuberous Roots.
Folk Uses : Tuberous roots eaten raw and often also as vegetable.
References : Arora & Pandey (1996), Rana *et al.* (2005).

Clematis cadmia Buch.-Ham. *ex* Wall.

Family : Ranunculaceae.

Distribution : Manipur (N.E. Himalaya).
Part Used : Roots.
Folk Use : Roots edible.
Reference : Sinha (1996).

Clinogyne dichotoma Salib.

Syn : *Donax arundinastrum* Laur.; *Phrynium dichotomum* Roxb.
Family : Marantaceae.

Hindi and Regional Names:

Hindi : Sitalpati;
Ass. : Murta;
Beng. : Muktapata, Sitalpaati.

Distribution : Assam in N.E. Himalaya. Bengal. Coromandal coast.

Part Used : Rhizomes.

Folk Uses : Rhizomes rich in starch and edible.

References : Agarwal (2003), Anonymous (1948-1976).

Codonopsis clematida (Chrenk) Clarke

Family : Campanulaceae.

Regional Name :

Kash. : Brutuge.

Distribution : Trans-N.W. Himalaya above 3,300m in J.&K.

Part Used : Roots.

Folk Uses : Roots consumed either raw or cooked by the locals, and also a source of edible flour.

References : Anonymous (1948-1976), Singh *et al.* (2001a).

Codonopsis ovata Benth.

Family : Campanulaceae.

Hindi Name :

Hindi : Luduti.

Distribution : N.W. Himalaya (J.&K. and H.P.): 2,800-4,000m.

Part Used : Large Fusiform Roots.

Folk Uses : Roots consumed by the locals either raw or cooked, and also its flour eaten in 'Lahaul' and other parts of N.W. Himalaya.

References : Ambasta (1986), Arora & Pandey (1996), Bennet *et al.*(1991), Rana *et al.* (2005), Samant *et al.* (2001), Singh *et al.* (2001a).

Coleus forskohlii Briq.

Family : Lamiaceae.

Hindi Name :

Hindi : Bhedi.

Distribution : Subtropical Himalaya. Kumaon. Deccan Peninsula. Gujarat. Bihar: 1,000-2,500 m.

Reproductive Cycle : November- December.

Part Used : Roots.

Folk Uses : Roots eaten raw or cooked as vegetable, and also pickled.

References : Agarwal (2003), Ambasta (1986), Arora (1985), Arora & Pandey (1996), Samant *et al.* (2001), Singh *et al.* (2001b).

Coleus parviflorus Benth.

Syn : *Coleus tuberosus* Benth.; *C. rotundifolius* Poir. Chev. & Perr.; *C. dysentericus* Baker.; *Germantia rotundifolia* Poir.

Family : Lamiaceae.

Hindi and Regional Names:

Hindi : Koorkan;
Tam. : Kooran Kizhangu.

Distribution : Cultivated mainly in M.P. and S. India.

Part Used : Tubers.

Folk Uses : Sweet, aromatic tubers used as a substitute for potato. Also used as tonic as they contain high content of minerals.

Active Constituents : Root tubers comprise of carbohydrates, 19.7; fats, 0.1; proteins, 1.3; minerals, 10.00; fibres, 0.4 and moisture, 77.6 %.

References : Agarwal (2003), Ambasta (1986), Anonymous (1948-1976), Arora (1985).

Coleus vettiveroides K.C.Jacob.

Family : Lamiaceae.

Hindi and Regional Names:

Hindi : Vettiver;
Tam. : Kuru-vettiver.

Distribution : Chiefly grown in S. India.
Part Used : Tubers.
Folk Use : Tubers eaten as substitute of potato.
References : Agarwal (2003), Anonymous (1948-1976).

Colocasia esculenta Linn. Schott. (Pl. 3C, D)

Syn : *C. antiquorum* Schott.; *Caladium esculentum* Vent.

Family : Araceae.

English, Hindi, Sanskrit and Regional Names:

Eng. : Cocoyam, Dasheen, Eddo-taro;
Hindi : Arvi, Ghuiya, Kachalu;
Sans. : Kaachu;
Ass. : Kaachu.
Beng. : Kaachu;
Kan. : Kachhchi, Shamagadde;
Ory. : Saru;
Mal. : Shembu;
Mar. : Alu;
Tam. : Seppen, Kizhangu;
Tel. : Chamadumpa.

Distribution : Tropical as well as subtropical Himalaya. Moist regions of India.
Reproductive Cycle : June-October.
Part Used : Tubers.
Folk Uses : Tubers consumed boiled or fried as vegetable. Also a source of 'Taro' flour which is used in soups, gravies, puddings, 'Taro' lactin and 'Taro' malt.
Active Constituents : Chief constituents of root tubers are carbohydrates, 22.1; fats, 0.1; minerals, 1.7 (Ca, 0.04%; P, 0.14%); moisture, 73.1; proteins 3.0%; vit. A, 40 I.U.; vit. B_1, 80 I.U. and traces of vit. C.

References : Agarwal (2003), Ambasta (1986), Anonymous (1948-1976), Arora (1981, 1985), Borthakur (1996), Jain (1981), Rana *et al.* (2005), Shortt (1887-1888).

Colocasia himalensis Royle

Family : Araceae.

Regional Name :

Kum. : Dhakol.

Distribution : W. Himalaya.

Part Used : Rhizomes.

Folk Uses : Rhizomes eaten cooked or after roasting in ashes.

References : Bhargava (1959), Gupta (1962).

Colocasia macrorrhiza Schott.

Family : Araceae.

Distribution : Tropical Himalaya.

Part Used : Corms.

Folk Uses : Corm eaten after proper washing and boiling.

References : Singh *et al.* (1999).

Remarks : Corms considered as poisonous and should be eaten only after proper washing and repeated boiling.

Colocasia nymphoefolia Kunth.

Family : Araceae.

Distribution : S. India. Madras.

Part Used : Tubers.

Folk Use : Tubers cooked and eaten.

Reference : Shortt (1887-1888).

Commelina benghalensis (Linn.) Forsk.

Family : Commelinaceae.

Regional Names :

Eng. : Black Caraway ;

Beng. : Kandhora;

Mar. : Kema;
Tam. : Kanaraghai.

Distribution : Common throughout the moist parts of India upto 2,000 m.
Reproductive Cycle : July-November.
Part Used : Rhizomes.
Folk Uses : Mucilaginous starchy rhizome cooked and used as vegetable. Also grounded into flour for making bread.
References : Agarwal (2003), Arora & Pandey (1996), Burkill (1935), Gammie (1902), Gupta (1962), Gupta & Kanodia (1968), Saxena (1979), Watt (1889-1896).

Coptis teeta Wall.

Family : Ranunculaceae.

English and Hindi Names:

Eng. : Gold Thread;
Hindi : Mameera, Teeta.

Distribution : E. Himalaya.
Part Used : Rhizomes.
Folk Uses : Rhizomes recommended for diabetic patients as food and also used as tonic.
Reference : Agarwal (2003).

Corydalis cashmiriana Royle

Family : Fumariaceae.

Distribution : N.W. Himalaya (J.&K. to Uttaranchal): 3,000-4,000 m.
Part Used : Tubers.
Folk Use : Root tubers edible.
Reference : Samant *et al.* (2001).

Costus speciosus (Koenig) J.E. Smith

Syn : *Banskia speciosa* Koeng apud Retz.;
Costus specious var. *nipalensis* (Rosc.) Baker.

Family : Costaceae.

Hindi, Sanskrit and Regional Names:

Hindi : Keu;
Sans. : Kemuka;

Beng. : Keul;
Mar. : Penva;
Tam. : Chengalvakoshtu.

Distribution : W. and N.E. Himalaya. S. India. Deccan plateau upto 1,200 m.
Reproductive Cycle : July- October.
Part Used : Rhizomes.
Folk Use : Rhizomes cooked and eaten.
References : Agarwal (2003), Ambasta (1986), Anonymous (1948-1976), Arora (1981), Arora & Pandey (1996), Aswal (2001), Bennet *et al.* (1991), Borthakur (1996), Jain (1981), Krishna *et al.* (1999), Mandal & Basu (1996), Pandey & Samant (2001), Sinha (1996).

Cotoneaster falconeri Kotz.

Family : Rosaceae.

Distribution : E. Himalaya (Arunanchal Pradesh): 2,500-3,200 m.
Part Used : Roots.
Folk Use : Roots edible.
Reference : Samant *et al.* (2001).

Crambe cordifolia Stevn.

Family : Brassicaceae.

Distribution : N. W. Himalaya.
Part Used : Roots.
Folk Use : Roots edible.
References : Ambasta (1986), Arora & Pandey (1996), Rana *et al.* (2005).

Crambe kotshyana Boiss.

Family : Brassicaceae.

Distribution : N.W. Himalaya (J. & K. and H.P.): 2,000-2,500m.
Part Used : Roots.

Folk Use : Roots used as vegetable by the locals.

Reference : Samant *et al.* (2001).

Crinum defixum Ker-Gawl.

Syn : *C. asiaticum* Roxb.

Family : Amaryllidaceae.

Hindi and Sanskrit Names:

Hindi : Pindar;
Sans. : Sukhadarshan.

Distribution : Throughout India.

Part Used : Roots.

Folk Use : Roots eaten during the time of scarcity.

References : Arora & Pandey (1996), Watt (1889-1896).

Curculigo orchioides Gaertn.

Family : Hypoxidaceae.

Hindi, Sanskrit and Regional Names:

Hindi : Kali musali;
Sans. : Mushali;
Beng. : Tala musli;
Kan. : Nelatati-gadde;
Tel. : Nelaty-gadde.

Distribution : Subtropical Himalaya from Kumaon eastwards and in W. Ghats from Konkan southwards upto 2,200 m.

Reproductive Cycle : August-September.

Part Used : Roots.

Folk Uses : The black coloured roots are a source of edible flour and also recommended as tonic.

Active Constituents : Roots rich in starch and fibres.

References : Agarwal (2003), Chandra (1997), Samant *et al.* (2001), Vartak (1981).

Curcuma amada Roxb.

Family : Zingiberaceae.

English, Hindi and Sanskrit Names:

Eng. : Mango-Ginger;
Hindi : Aan Haldi;
Sans. : Karpura Haridra.

Distribution : Humid tracts of S. India.

Part Used : Rhizomes.

Folk Uses : Rhizomes used as condiment and considered akin to fresh mangoes. Also used in pickles and sauces.

Active Constituents : Chief pungent principle of rhizomes is due to d-á-pinene, d-camphor, 1-â-curcumene as hydrochloride, d-curcumene and phytosterol.

References : Ambasta (1986), Arora & Pandey (1996), Chevallier (1996), Jain (1995), Rastogi & Mehrotra (1990), Tripathi & Goyal (2001), Watt (1889-1896).

Curcuma angustifolia Roxb.

Family : Zingiberaceae.

English , Hindi, Sanskrit and Regional Names:

Eng. : East Indian Arrowroot, Narrow Leaved Turmeric, Wild Arrowroot;
Hindi : Tavakhira, Tokhur.
Sans. : Gavayodbhava, Pishtika, Tavakshir;
Mar. : Taavakhira;
Tam. : Arrarut, Kizhangu;
Tel. : Araut -gadde.

Distribution : Tropical and C. Himalaya. N. Bengal. W. Bihar. Bombay. S. India upto 1,800 m.

Reproductive Cycle : July- October.

Part Used : Tuberous Rhizomes.

Folk Uses : Tubers yield fragrant starch resembling arrowroot which is easily digestible and recommended for invalid children. Rhizomes also used as condiment in confectionery and drinks to give a calm and cooling effect.

Active Constituents : Tuberous rhizomes comprise of ash, 4.46; carbohydrates (starch), 26.00-46.00; fats, 0.2; fibres, 8.42; moisture, 13.13 and proteins 30.73 %.

References : Ambasta (1986), Anonymous (1948-1976), Arora (1981, 1985), Aswal (2001), Pandey & Samant (2001), Bennet *et al.* (1991), Jain (1995) Sahu (1996), Tripathi & Goyal (2001), Watt (1889-1896).

Curcuma aromatica Salisb.

Syn : *C. zedoaria* Roxb.
Family : Zingiberaceae.

English, Hindi & Sanskrit Names:

Eng. : Cochin Turmeric, Wild Turmeric, Yellow Zedoary;
Hindi : Ban-haldi, Banharidra, Janglihaldi;
Sans. : Aranyaharidra, Sholi, Van-haridra.

Distribution : Throughout India chiefly Bengal, W. Peninsula (semicultivated).

Reproductive Cycle : May-September.

Part Used : Rhizomes.

Folk Uses : Rhizomes used as spice and condiment for vegetables and pickles, and also as tonic and for colouring edible dishes.

Active Constituents : Colour and fragrance of essential oil are due to xanthorrhizol (25.7), â-curcumene (25.5) and ar-curcumene (18.6%).

References : Ambasta (1986), Arora (1981), Aswal (2001), Chopra *et al.* (1956), Pandey & Samant (2001), Jain (1995), Kirtikar & Basu (1984), Rastogi & Mehrotra (1993, 1995), Tripathi & Goyal (2001), Watt (1889-1896).

Curcuma caesia Roxb.

Family : Zingiberaceae.

English, Hindi and Regional Names:

Eng. : Black Zedoary;
Hindi : Nar-Kachura;
Beng. : Kali Haldi;
Tel. : Manupusupu.

Distribution : Common in Bengal.

Part Used : Rhizomes.

Folk Use : Highly aromatic rhizomes used for flavouring dishes.

Active Constituents : Rhizomes contain albuminoids, 22.94; ash, 7.57; carbohydrates (starch), 18.75; essential oils and resin, 4.37; fats, 0.2; fibres, 25.20; gum and organic acids, 10.10; moisture, 9.76%. Essential oils responsible for the flavour and fragrance are bornenylinecamphene, cineole, curcumine, linalool and zingiberol.

References : Agarwal (2003), Ambasta (1986), Anonymous (1948-1976) Arora (1981), Rastogi & Mehrotra (1995), Tripathi & Goyal (2001), Watt (1889-1896).

Curcuma caulina J.Grah.

Syn : *Hitchenia caulina* Grah.

Family : Zingiberaceae.

English, Hindi and Regional Names:

Eng. : Indian Arrowroot;
Hindi : Tikhur ;
Bomb. : Chavara, Chovera.

Distribution : Deccan. Considered as substitute for the arrowroot and consumed as food after proper washing. Also employed as flavouring agent in confectionery.

Active Constituents : Tubers rich in starch 10.9-18.3 % of fresh weight.

References : Agarwal (2003), Ambasta (1986), Arora & Pandey (1996), Bennet *et al.* (1991), Watt (1889-1896).

Curcuma domestica (Medik.) Valeton (Pl. 4B, C)

Syn : *C. longa* Linn.
Family : Zingiberaceae.

English, Hindi, Sanskrit and Regional Names:

Eng. : Indian Saffron, Turmeric;
Hindi : Halada, Haldi, Hardi;
Sans. : Aneshta, Bahula, Gauri, Haridra, Nisa, Rajani, Ralri;
Kan. : Arishna;
Tam. : Manjal;
Tel. : Pushpu.

Distribution : Cultivated throughout India.

Reproductive Cycle : June-November.

Part Used : Rhizomes.

Folk Uses : Aromatic rhizomes used as a condiment and colouring agent in a number of Indian dishes.

Active Constituents : Rhizome comprises of carbohydrates, 26.00; fats, 5.1; minerals, 3.5%, moisture, 13.10; proteins, 6.3; and cellulose, etc. Major colouring matter is curcumin. Chief constituents of turmeric oil responsible for the flavour are, á- and â- turmerones, zingiberene, á- and â-pinenes, camphene, limonene, terpinene, caryophyllene, curcumene, linalool, turmerone, ar-turmerone, borneol, isoborneol, camphor, eugenol, cineole, curdione, curzerenone, curcumin, monodemethoxy-curcumin and didemethoxy-cucurcumin.

References : Agarwal (2003), Ambasta (1986), Anonymous (1948-1976), Arora (1981), Aswal (2001), Bennet (1987), Bhattacharjee (1998), Chevallier (1996),

Jain (1995), Mehtani & Ingle (1999), Kirtikar & Basu (1984), Rastogi & Mehrotra (1991,1993,1995), Sharma & Singh (2001), Tripathi & Goyal (2001).

Curcuma leucorhiza Roxb.

Family : Zingiberaceae.

Hindi Name :

Hindi : Tikar.

Distribution : E. India. Bihar. W.Bengal.

Part Used : Rhizomes.

Folk Uses : Rhizomes yield starch like that of arrowroot and used by locals in food preparation.

References : Ambasta (1986), Arora & Pandey (1996).

Curcuma montana Rosc.

Family : Zingiberaceae.

Distribution : Throughout India.

Reproductive Cycle : September- October.

Part Used : Rootstocks.

Folk Use : Rootstocks edible.

Reference : Bennet *et al.* (1991).

Curcuma neilgherrensis Roxb.

Family : Zingiberaceae.

Distribution : Maharashtra. Goa.

Part Used : Rhizomes.

Folk Use : Rhizomes edible.

Reference : Vartak (1981).

Curcuma pseudomontana Grah.

Family : Zingiberaceae.

Regional Names :

Kon. : Sinderwar, Sinderwani;
Mar : Kachora, Sindalvan.

Distribution : Maharashtra. Goa.
Reproductive Cycle : June- November.
Part Used : Rhizomes.
Folk Use : Rhizomes yield a form of arrowroot which is cooked and eaten.
References : Ambasta (1986), Chandra (1997), Datar & Vartak (1975), Watt (1889-1896).

Curcuma rubescenses Roxb.

Family : Zingiberaceae.

Distribution : N.E. Himalaya.
Part Used : Rhizome.
Folk Use : Rhizome yield a form of arrowroot starch which is cooked and eaten.
Reference : Ambasta (1986).

Curcuma zedoaria Rosc.

Syn : *Amomum zedoaria* Christ; *C. zerumbet* Roxb.
Family : Zingiberaceae.

English , Hindi and Sanskrit Names:

Eng. : Long Zedoary, Round Zedoary; Zeodary;
Hindi : Kachura, Kalihaldi;
Sans. : Dravida, Gandhasara, Jatala, Kalpaka, Mukhya, Sati, Vedhya.

Distribution : Wild in E. Himalaya. Cultivated more or less throughout India upto 1,500 m.
Reproductive Cycle : April – June.
Part Used : Rhizomes.
Folk Uses : Starchy rhizome is a source of 'Zedoria' or 'Zeodary' which is boiled and eaten. Also it finds use as condiment and spices.
Active Constituents : Rhizome comprises of carbohydrates (starch), 82.6 (major is amylase); moisture, 13.1; proteins 0.5 and fats, 0.2. Essential oils isolated are pyrocurzerenone, guaiane-zedoarondiol, curcumenone,

curcumanolides zedoarol, germacrone, curzeone, furanogermenone and curcumenol.

References : Agarwal (2003), Ambasta (1986), Arora (1981), Aswal (2001) Chevallier (1996), Kirtikar & Basu (1984), Pandey & Samant (2001), Rastogi & Mehrotra (1990, 1991, 1995), Tripathi & Goyal (2001), Vartak (1981), Watt (1889-1896).

Cycas revoluta Thunb.

Family : Cycadaceae.

Hindi Name :

Hindi : Madangameshvari.

Distribution : Cultivated in Indian gardens as ornamental plant.

Part Used : Roots.

Folk Use : 'Sago' starch extracted from root pith is used during the time of food scarcity and famine.

References : Agarwal (1986), Ambasta (1986), Anonymous (1948-1976).

Cyanotis vaga (Lour.) Schult.

Family : Commelinaceae.

Distribution : N.W. Himalaya (Kumaon and Garhwal): 1,000- 2,000 m.

Part Used : Tender roots.

Folk Uses : Tender roots edible.

Reference : Samant *et al.* (2001).

Cyperus bulbosus Vahl.

Family : Cyperaceae.

Hindi and Regional Names:

Hindi : Gaddi- kans, Motha;
Tam. : Shilandi;
Tel. : Puri -dumpa, Puri-gaddi.

Distribution : Throughout Gangetic plains. Coastal Indian belt of Peninsular India.

Part Used : Small Tuberous Bulbs.

Folk Uses : Dried and pulverized tuberous bulbs having flavour of potato are baked into bread after mixing with jowari, bajri, (millet flour) or wheat flour. Also roasted bulbs relished by the 'Bhils'.

Active Constituents : Chemical composition of bulbs reveals (grams /100g) ash, 1.58g.; carbohydrates (soluble), 82.46g; fats, 0.58g; fibres, 1.51g; proteins, 2.27g and water, 11.60g.

References : Agarwal (2003), Ambasta (1986), Arora & Pandey (1996), Rastogi & Mehrotra (1995), Gammie (1902), Gupta & Kanodia (1968), Rana *et al.* (2005), Saxena (1979).

Cyperus esculentus Linn.

Family : Cyperaceae.

English, Hindi, Sanskrit and Regional Names:

Eng. : Chuffa,, Rush- nut Tiger- nut;
Hindi : Chichoda, Chufa;
Sans. : Kaseruka;
Mani. : Kaotham.

Distribution : W. Ghat. Nilgiri Hills.

Part Used : Tubers.

Folk Uses : Tubers eaten raw or cooked after boiling and roasting.

Active Constituents : Root tubers rich in albuminoids, 5.21; digestible carbohydrates (not starch), 24.79; fibres, 5.83, minerals, 1.48; moisture, 14.15; and oils, 25.82 %.

References : Ambasta (1986), Arora & Pandey (1996), Anonymous (1948-1976), Singh *et al.* (2001b), Rana *et al.* (2005).

Cyperus metzii (Hochst.) Mattf. & Kukenthal

Syn : *Kyllingia squamata* Vahl.

Family : Cyperaceae.

Distribution : W. Himalaya. W. Ghats.

Part Used : Rhizomes.

Folk Use : Juice of the rhizomes used as a flavouring agent for foods and medicine.

References : Ambasta (1986), Mehtani & Ingle (1999).

Cyperus rotundus Linn.

Syn : *C. hexastachyos* Roxb.; *C. retzii* Nees. non Poir.; *C. tubersosus* Sensu Cl. non Roxb.

Family : Cyperaceae.

English, Hindi, Sanskrit and Regional Names:

Eng. : Nut Grass ;
Hindi : Motha, Mutha ;
Sans. : Abda, Bhadrakshi, Musta, Mustaka ;
Mar. : Kavala –Motha.

Distribution : Kashmir to Shimla. Garhwal. Khasia Hills. C. Himalaya. Maharashtra. Goa. Warmer regions of India upto 1,300 m.

Part Used : Dried Tuberous Roots.

Folk Uses : Tubers used as a substitute for the 'arecanut' in Sikkim and recommended as famine food. These can be eaten raw or cooked with spices and its flour used for making bread.

References : Agarwal (2003), Ambasta (1986), Bennet (1987), Das *et al.* (1996), Kirtikar & Basu (1984), Kshirsagar *et al.* (2003), Raju (2000), Rastogi & Mehrotra (1990) Samant *et al.* (2001), Saxena (1979), Singh & Pandey (1998), Watt (1889-1896).

Cyperus scariosus R.Br.

Family : Cyperaceae.

Hindi and Sanskrit Names:

Hindi : Nargmutha;
Sans. : Chakranksha, Kalapini, Nagarotha, Nargmusta.

Distribution : Damp places in Bengal, Panjab, U.P., Bengal, S. India.

Part Used : Tubers.

Folk Uses : Tubers used as spice and condiment.

References : Agarwal (1986), Ambasta (1986), Kirtikar & Basu (1984), Rastogi & Mehrotra (1990), Watt (1889-1896).

Dactylorhiza hatagirea D. Don.

Family : Orchidaceae.

Distribution : Trans-N.W. Himalaya (J. & K. and H.P.) N.W. Kumaon and Garhwal regions of Himalaya: 2,800 - 4,000 m.

Part Used : Tubers.

Folk Use : Tubers eaten cooked or raw.

Reference : Samant *et al.* (2001).

Daucus carota Linn. var. sativ*a*

Family : Apiaceae.

English, Hindi, Sanskrit and Regional Names:

Eng. : The carrot;
Hindi : Gajar;
Sans. : Gajar, Jajar;
Beng. : Mor-muj.

Distribution : Cultivated throughout India for its flerhy roots.

Part Used : Roots

Folk Uses : Flerhy tap roots used as salad or cooked as vegetable. It is used for pickling, soups and also in curries and pies, etc. canned. Carrot juice is a rich source of carotene (vit. A) and also used to colour the butter and other food items.

Active Constituents : Carrot root gives crystalised & uncrystalised sugars, a little starch, extractive glutin, albumins, volatile oils, vegetable jelly or pectins, malic acid, saline matter, lignin, a,b,g carotene ($C_{18}H_{30}O$), vit. A, 4,000 I.U.; and vit. B complex, 1-60 I.U. Red variety is rich in iron with 10% fibres whereas yellow variety rich in sulphur but low in fibres.

References : Agarwal (2003), Ambasta (1986), Anonymous (1948-1976), Arora & Pandey (1996), Watt (1889-1896).

Decalepis hamiltonii Wight & Arn.

Family : Asclepiadaceae.

Hindi and Regional Names:

Hindi : Mahali;
Ass. : Rui-kaulang;
Kan. : Makalibru;
Tam. : Mahli- Kizhangu.

Distribution : N.E. Himalaya. Common in W. Ghats. Deccan Peninsula.

Part Used : Roots.

Folk Uses : Aromatic roots used as spice and condiments. Also pickled with lemon or eaten as such.

Active Constituents : The volatile principle is due to 4-0-methyl resor-cycaldehyde.

References : Agarwal (2003), Ambasta (1986), Anonymous (1948-1976), Arora & Pandey (1996), Bennet *et al.* (1991), Chandra (1997), Chopra *et al.* (1956), Jain (1996), Mittre (1981), Sahu (1996).

Dicentra thalictrifolia Hk.f.

Family : Papaveraceae.

Distribution : E. Himalaya (Arunanchal Pradesh): 1,500-2,500m.

Part Used : Tubers.

Folk Use : Tubers edible.

References : Anonymous (1948-1976), Samant *et al.* (2001).

Didymospermum namum H.Wendl. & Drude.

Family : Arcaceae.

Distribution : Arunanchal Pradesh: 300-1,200 m.

Part Used : Bulbs.

Folk Use : Underground bulbs alongwith the tender shoots are edible.

Reference : Samant *et al.* (2001).

Dioscorea aculeata Linn.

Family : Dioscoreaceae.

English, Hindi, Sanskrit and Regional Names:

Eng. : Goa -Potato, Guina -Yam, Kawi -Yam, Prickly-stemed Yam;
Hindi : Man -alu, Mou-alu, Pindalu;
Sans. : Bir-Sangi;
Beng. : Kanta-aluu;
Mal. : *Kahu;*
Tam. : Kantu-Kalangu;
Tel. : Genaru.

Distribution : W. Himalaya (Kumaon). C. Bengal. Semicultivated in W. and S. India.

Part Used : Tubers.

Folk Uses : Delicate, white tubers rich in starch are eaten raw or cooked.

References : Ambasta (1986) Anonymous (1948-1976), Arora & Pandey (1996), Samant *et al.* (2001), Bhargava (1981), Watt (1889-1896).

Dioscorea alata Linn.

Syn : *D. atropurpurea* Roxb.; *D. globosa* Roxb.; *Dioscorea purpurea* Roxb.; *Dioscorea rubella* Roxb.

Family : Dioscoreaceae.

English, Sanskrit ,Hindi and Regional Names:

Eng. : Greater Yam, True Asiatic Yam;
Hindi : Chupri Alu, Khamalu;
Sans. : Pindalu;
Ass. : Kattalu;
Beng. : Chupri Alu;
Mal. : Kavathu;

Tam. : Perum -valli;
Tel. : Pendalamu.

Distribution : Cultivated practically all over India. Also wild in Himalaya upto 2,900 m.

Reproductive Cycle : October- December.

Part Used : Tubers.

Folk Uses : Tubers when boiled and mixed with traditional spices make excellent vegetables and eaten by N.E. tribals. 'Yam' with purple tint is used for colouring the ice creams.

Active Constituents : Tubers rich in albuminoids, 7.96-15.68; ash, 4.23-7.28; carbohydrates, 71.67-85.2; fats, 0.59-1.26; fibres 2.19-6.12 and P, 0.44-0.95%. Out of total carbohydrates, 21% is starch on an average.

References : Ambasta (1986), Anonymous (1948-1976), Arora (1981), Arora & Pandey (1996), Das & Chattopadhyay(2001), Lalramnghinglova (2002), Samant et al. (2001), Shortt (1887-1888), Sinha (1996), Watt (1889-1896).

Dioscorea anguina Roxb.

Family : Dioscoreaceae.

Regional Name :

Beng. : Kukur Alu.

Distribution : N.W. Himalaya. Kolkatta.

Reproductive Cycle : August-December.

Part Used : Tubers.

Folk Uses : Tubers eaten by the poor people when hard pressed or at the time of emergency.

References : Krishna *et al.* (1999), Watt (1889-1896).

Dioscorea arachinida Prain *ex* Burk.

Family : Dioscoreaceae.

Regional Name :

Ass. : Tinnipotialu.

Distribution : N.E. Himalaya (Assam) upto 1,400 m.

Part Used : Tubers.

Folk Uses : Tubers used as vegetable by the local tribals.

References : Anonymous (1948-1976), Arora & Pandey (1996), Borthakur (1996), Samant *et al.* (2001), Watt (1889-1896).

Dioscorea atropurpurea Roxb.

Family : Dioscoreaceae.

English Name :

Eng. : Dark purple Yam, Malliaca Yam, Rangoon Yam.

Distribution : N.E Himalaya.

Reproductive Cycle : July – December.

Part Used : Tubers.

Folk Uses : Deep-purple tubers eaten as vegetable by the locals and sold as 'Rangoon Yam' in the markets.

References : Watt (1889-1896).

Dioscorea batatus Decaisne.

Family : Dioscoreaceae.

Distribution : N.E. Himalaya.

Part Used : Tubers.

Folk Use : Used as vegetable by the locals.

Reference : Watt. (1889-1896).

Dioscorea belophylla (Prain) Voight. *ex* Haines (Pl. 5A, B)

Syn : *D. glabra* sensu Hook.f.

Family : Dioscoreaceae.

Regional Names:

H.P. : Tardi;

Pub. : Zamin Kand;

Sant. : Mutur;

Tel. : Naratega.

Distribution : N.W. Himalaya, and C. Himalaya: 600-1,500 m. Almost throughout India.

Reproductive Cycle : September- February.

Part Used : Tubers.

Folk Uses : Tubers eaten raw or cooked as vegetable and also pickled.

References : Ambasta (1986), Anonymous (1948-1976), Arora & Pandey (1996), Bennet *et al.* (1991), Jain (1996), Rana *et al.* (2005), Samant *et al.* (2001), Singh (1996), Sinha (1996), Watt (1889-1896).

Dioscorea bulbifera Linn.

Syn : *Dioscorea crispata* Roxb.; *D. pulchella* Roxb.; *D. sativa* Thunb. non Linn.; *D. vesicolor* Buch. *ex* Wall.

Family : Dioscoreaceae.

English, Hindi and Regional Names:

Eng. : Bulb bearing Yam;
Hindi : Zamin -kand;
Ass. : Kath -alu;
Mar. : Kadu- Karanda.

Distribution : N.E. and subtropical Himalaya and common in almost all parts of India upto 1,400 m.

Reproductive Cycle : July – January.

Part Used : Tubers.

Folk Uses : Starchy tubers boiled and eaten at the time of scarcity. Also these are cooked in curries after roasting or frying, and are also pickled. In Assam the tubers are boiled in acidic water before being eaten. For pickling purpose, small pieces of tubers are put into vinegar or in the mixture of mustard oil with salt.

Active Constituents : Tubers (dry matter basis) comprise albuminoids, 7.36-13.31; ash, 3.31-7.08; carbohydrates, 75.11-81.39; fats, 0.75-1.28; fibres, 3.28-9.64 and P, 0.45-0.77%.

References : Ambasta (1986), Anonymous (1948-1976), Arora (1981, 1985), Arora & Pandey

(1996), Bennet *et al.* (1991), Borthakur (1996), Das & Chattopadhyay (2001), Jain (1996), Krishna *et al.* (1999), Mandal & Basu (1996), Rana *et al.* (2005), Raju (2000), Rao & Henery (1996), Samant *et al.* (2001), Saini (1996), Singh (1996), Singh & Anand (1996), Sinha (1996), Vartak (1981),Watt (1889-1896).

Dioscorea daemona Roxb.

Syn : *Helmia daemona* Kunth.; *Ubium silvestre* Rumph.
Family : Dioscoreaceae.

Regional Name :

Tel. : Pudumpa, Puli-Dumpa.

Distribution : N.W. Himalaya (Kumaon). S. India (Karnataka, & Tamilnadu).

Part Used : Tubers.

Folk Use : Tubers very nauseous and should be eaten after proper boiling.

Reference : Watt (1889-1896).

Dioscorea deltiodea Wall. *ex* Grisb.

Family : Dioscoreaceae.

Regional Names :

Kash. : Kildri, kins, Kithi, Krish;
Pub. : Kitra, Kniss Kris, Tar.

Destribution : N.W.Himalaya.

Part Used : Tuberous Bulbs.

Folk Uses : Tuberous bulbs edible after boiling.

Active Constituents : Tubers (dry matter basis) comprise of albuminoids, 10.82; ash, 9.65; carbohydrates, 71.29; fats, 1.72; fibres, 6.52 and P, 0.94 %.

References : Arora & Pandey (1996), Rana *et al.* (2005), Samant *et al.* (2001), Watt (1889-1896).

Dioscorea esculenta (Lour.) Burkil.

Syn : *D. aculeata* Linn.; *D. fasciculata* Roxb.; *D.spinosa* Roxb. *ex* Wall.

Family : Dioscoreaceae.

English, Hindi and Regional Names:

Eng. : Karen Potato, Lesser Yam;
Hindi : Suthini;
Beng. : Susni Alu , Suthini;
Mal. : Mukkizhangu ,Mullu- kilangu;
Tel. : Sila Kadom.

Distribution : Throughout India and Andaman Islands.

Part Used : Tubers.

Folk Uses : Starchy tubers consumed in boiled or roasted form.

Active Constituents : Tubers comprise of albuminoids, 12.02; ash, 6.65; carbohydrates, 77.29; fats, 1.02; fibres, 5.52 and P, 0.94 % on dry matter basis.

References : Agarwal (2003), Ambasta (1986), Anonymous (1948-1976), Arora & Pandey (1996) Bennet *et al.* (1991), Jain (1981), Radhakrishan *et al.* (1996), Watt (1889-1896).

Dioscorea fasciculata Roxb.

Family : Dioscoreaceae.

Regional Name :.

Beng. : Sadhhui.

Distribution : All over India cultivated as well as wild particularly in N.E. Himalaya.

Part Used : Root Tubers.

Folk Uses : Roots consisting of small, several, light coloured tubers used as food by the natives and for the manufacturing the starch. Also the flesh of tuber is sweet and considered tasty

Reference : Watt (1889-1896).

Dioscorea floribunda Mart. & Gal.

Family : Dioscoreaceae.

Distribution : E. Himalaya: 400-1,500 m.
Part Used : Root Tubers.
Folk Use : Root tubers edible.
Reference : Samant *et al.* (2001).

Dioscorea glabra Roxb.

Family : Dioscoreaceae.

Regional Names :

Ass. : Rui-ding;
Mani. : Ha;
Tel. : Eddu-toka, Dumpa, Nag Tega.

Distribution : N. E. Himalaya upto 2,500 m. Found all over India and in Andaman and Nicobar islands, often cultivated.
Reproductive Cycle : June-February.
Part Used : Tubers.
Folk Uses : Tubers delicious but not much relished as they become gluey on cooking. However, tubers of 'Khasia' and 'Garo' hills are better in quality and often baked into bread.
Active Constituents : Various constituents of tuber are albuminoids, 9.73-10.13; ash, 5.19-6.70; carbohydrates, 77.79-78.23; fats, 1.29-1.42; fibres, 3.92-5.00 and P, 0.58-0.64 % on dry matter basis.
References : Ambasta (1986), Anonymous (1948-1976), Arora & Pandey (1996) Borthakur (1996), Das & Chattopadhyay (2001), Rana *et al.* (2005), Rao & Henery (1996), Samant *et al.* (2001), Sinha (1996), Watt (1889-1896).

Dioscorea hamiltonni Hook.f.

Syn : *Dioscorea hookeri* Prain.

Family : Dioscoreaceae.

Regional Name:

M.P. : Nagar Kanda.

Distribution : C. India (M.P.) and N.E. Himalaya: 1,500-2,500 m.

Reproductive Cycle : September- December.

Part Used : Tubers.

Folk Uses : Tubers delicious and most esteemed in Kerala and among the 'Mirkis' tribals of Assam.

Active Constituents : Tubers (dry matter bases) are rich in albuminoids, 8.30; ash, 3.91; carbohydrates, 85.50; fats, 0.77; fibres, 1.52 and P, 0.55 %.

References : Ambasta (1986), Anonymous (1948-1976), Arora & Pandey (1996), Borthakur (1996), Jain (1981), Samant *et al.* (2001), Sahu (1996), Watt (1889-1896).

Dioscorea hirusta Dennst.

Family : Dioscoreaceae.

Distribution : N.E. Himalaya. Tripura.

Part Used : Tubers.

Folk Use : Tubers edible among the tribals of Tripura.

Reference : Singh *et al.* (1999).

Dioscorea hispida Dennst. (Pl. 6A)

Syn : *D. daemona* Roxb.; *D. hirusta* Dennst.

Family : Dioscoreaceae.

Hindi and Regional Names:

Hindi : Karukandu;
M.P. : Baichandi-Kanda,Kalia Papad;
Mar. : Baichandi;
Tam. : Perperndai;
Tel. : Tellagingigeddalu.

Distribution : Found almost throughout India. N.E. and N.W. Himalaya: 1,000-1,800 m.

Reproductive Cycle : July- November.

Part Used : Tubers.

Folk Uses : Tubers very large (upto 35 Kg), acrid and poisonous but used as food at the time of scarcity by the 'Lepachas' after soaking in running water stream for about a week before cooking. Tribals of 'Bastar' district on the other hand use them by first dipping in water overnight and then made these into chips or 'Papad' and eaten after frying.

Active Constituents : Tubers (dry matter bases) reveal the occurrence of albuminoids, 7.20-9.12; ash, 4.05-4.61; carbohydrates, 81.45-81.89; fats, 0.97-1.10; fibres, 3.28-6.28; and P, 0.52-0.70% .

References : Ambasta (1986), Anonymous (1948-1976), Arora (1981), Arora & Pandey (1996), Bennet *et al.* (1991), Jain (1981), Krishna *et al.* (1999), Rao & Henery (1996), Sahu (1996), Samant *et al.* (2001), Watt (1889-1896).

Dioscorea kumaonesis Kunth.

Family : Dioscoreaceae.

Distribution : C. and N.W. Himalaya (Kumaon and Garhwal) upto 1,800 m.

Part Used : Tubers.

Folk Uses : Tubers eaten boiled or cooked by the locals.

References : Rana *et al.* (2005), Watt (1889-1896).

Dioscorea lepacharum Prain & Burkil.

Family : Dioscoreaceae.

Distribution : N.E. and N.W. Himalaya.

Part Used : Tubers.

Folk Uses : Starchy tubers eaten boiled as vegetable.

Reference : Borthakur (1996).

Dioscorea melanophyrma Prain & Burkill

Syn : *D. tenuii* Kunth.; *D. kumaonensis* Auct. non Kunth.

Family : Dioscoreaceae.

Hindi Name :

Hindi : Mag, Manghaii.

Distribution : N.E. and N.W.Himalaya from Uttaranchal to Assam: 600-3,000 m.

Reproductive Cycle : September- December.

Part Used : Tubers.

Folk Use : Tubers edible after boiling.

References : Anonymous (1948-1976), Arora & Pandey (1996), Bennet *et al.* (1991) Samant *et al.* (2001), Watt (1889-1896).

Dioscorea nummularia Lam.

Family : Dioscoreaceae.

Regional Name :

Tel. : Karu pendalam.

Distribution : W. Bengal.

Reproductive Cycle : August-February (Next year).

Part Used : Tubers.

Folk Use : Tubes eaten as vegetable only after proper boiling.

Reference : Watt (1889-1896).

Dioscorea oppositifolia Linn.

Family : Dioscoreaceae.

Hindi and Regional Names:

Hindi : Kanta -Alu;
Kerala : Kavalai, Kavalan;
Mal. : Karanda;
Mar. : Passpoli;
Ori : Pani Alu
Tam. : Kavalu;
Tel. : Avatengategge.

Distribution : E. India. W. Ghats. Subhimalayan tract.

Reproductive Cycle : August- November.
Part Used : Tuberous Roots.
Folk Use : Boiled tubers eaten after removing its skin. Considered a great delicacy by the forest tribals.
Active Constituents : Tuber (dry matter bases) contains albuminoids, 14.70; carbohydrates, 68.54; fats, 1.5-2.0; fibres, 6.55; P, 0.71 % and ash.
References : Anonymous (1948-1976), Arora & Pandey (1996) Singh *et al.* (1999), Bennet *et al.* (1991), Das & Chatopadhyay (2001), Radhakrishanan *et al.* (1996), Rao & Henery (1996), Saini (1996), Samant *et al.* (2001), Vartak (1981),Watt (1889-1896).

Dioscorea pentaphylla (Linn.) Willd. (Pl. 6B)

Syn : *D. jacquemontii* Hook.f.; *D. triphylla* Linn.
Family : Dioscoreaceae.

English, Hindi and Regional Names:

Eng. : The Kawan Yam;
Hindi : Bhusa, Gazaria, Kanta Alu;
Beng. : Suralu ;
M.P. : Loc- Konge;
Mal. : Ulsi;
Man. : Dragal;
Mani. : Ha;
Tam. : Kattukilangu;
Tel. : Duk-Apen- Dalamu.

Distribution : N. W. Himalaya, tropical Himalaya and almost all over India (wild as well as cultivated): 1,000- 1,300m.
Reproductive Cycle : May –February (Next Year).
Part Used : Tubers.
Folk Uses : Tubers eaten with salt after keeping in a running stream for one night and one day, and then boiled repeatedly for five to six time to remove the rind. Those collected during rains are sun-dried and can be used throughout the year as vegetable.

Active Constituents : On dry matter basis chief constituents of the tubers are albuminoids, 7.96-15.18; ash, 4.23-7.28; carbohydrates, 71.67-85.02; fats, 0.59-1.26; fibres, 2.19-6.12 and P, 0.44-0.98%.

References : Abraham (1981),Agarwal (2003), Ambasta (1986) Anonymous (1948-1976), Arora & Pandey (1996), Bennet *et al.* (1991),Bhargava (1959,1981), Borthakur (1996),Das & Chatopadhyay (2001), Gammie (1902), Gupta (1962), Gupta & Kanodia (1968), Jain (1981) Mandal & Basu (1996), Patole & Jain (2002), Radhakrishan *et al.* (1996), Rana (2004), Rana et al. (2005), Rao & Henery (1996), Samant *et al.* (2001), Singh (1996), Singh & Anand (1996), Sinha (1996), Vartak (1981), Watt (1889-1896).

Remarks : Tubers should be consumed only after repeated boiling as they are acrid and poisonous when raw.

Dioscorea prazeri Prain & Burkil.

Syn : *D. clarkei* Prain & Burkil.; *D. deltoida* Wall. var. *sikkimensis* Prain.; *D. sikkimensis* Prain & Burkil.

Family : Dioscoreaceae.

Distribution : C. and N.W. Himalaya (J. & K. and H.P.): 1,000-1,600 m.

Parts Used : Tuberous Bulbs.

Folk Uses : Tuberous bulbs eaten as vegetable by the locals.

References : Ambasta (1986), Anonymous (1948-1976), Arora & Pandey (1996), Samant *et al.* (2001), Watt (1889-1896).

Remarks : Certain varieties are used as fish poisons and should be eaten carefully as considered poisonous.

Dioscorea puber Blume.

Syn : *D. anguina* Roxb.
Family : Dioscoreaceae.

Regional Name :

Ass. : Rui-Han;
Beng : Kakur-Alu.

Distribution : C. and E. Himalaya. Kerala. Mostly in wet places: 900-1,500 m.

Reproductive Cycle : September-January.

Part Used : Tuberous Bulbs.

Folk Use : Tuberous bulbs eaten as vegetable by the locals.

Active Constituents : Tuberous bulbs contain albuminoids, 11.44-12.45; ash; 3.72-4.54; carbohydrates, 78.42-81.34; fats,0.56-1.13; fibres, 2.94-3.36 and P, 0.52-0.57 %.

References : Agarwal (2003), Ambasta (1986), Anonymous (1948-1976), Arora & Pandey (1996), Jain (1981), Rao & Henery (1996), Samant *et al.* (2001), Watt (1889-1896).

Dioscorea purpurea Roxb.

Family : Dioscoreaceae.

Distribution : W.Bengal.

Part Used : Tubers.

Folk Use : Tubers considered as third best among the 'Yams'. Dark purple coloured tubers used as vegetable.

Reference : Watt (1889-1896).

Dioscorea quinata (J.F.Geml.) Wall.

Family : Dioscoreaceae.

Regional Name :

Kum. : Majiya, Muniya.

Distribution : Kumaon (N.W.Himalaya).

Parts Used : Tubers.

Folk Use : Pieces of tubers steeped in water and boiled prior to eating.

References : Bhargava (1981), Rana *et al.* (2005), Watt (1889-1896).

Dioscorea rubella Roxb.

Family : Dioscoreaceae.

Regional Name :

Beng. : Goran Alu, Guraniya-Alu.

Distribution : W.Bengal.

Part Used : Tubers.

Folk Uses : Tubers- an excellent 'Yam' are fleshy and farinaceous, and used as an excellent substitute for the potato.

Reference : Watt (1889-1896).

Dioscorea sagittata Royle.

Family : Dioscoreaceae.

Regional Name :

Kum. : Tageer, Tareer, Tair.

Distribution : N.W. Himalaya (Kumaon region).

Folk Use : Axillary tubers steeped in water and boiled prior to eating.

Reference : Watt (1889-1896).

Dioscorea sativa Linn. Kunth.

Family : Dioscoreaceae.

English, Hindi and Regional Names:

Eng. : CommonYam;
Hindi : Ratalu;
Bomb. : Kadia Kand, Mano Kand, Vaj Kand.

Distribution : Cultivated in greater parts of India.

Part Used : Tubers.

Folk Uses : Pieces of tubers boiled and then leached in flowing water to remove the toxic elements, and then mixed with 'konda' or some other flour prior to making bread. Roasted tubers also relished.

References : Gammie (1902), Watt (1889-1896).

Remarks : Tubers acrid and toxic and be boiled repeatedly before eating.

Dioscorea tomentosa Koenig.

Syn : *Helmia tomentosa* Kunth.

Family : Dioscoreaceae.

Distribution : C. India. E. Ghats.

Part Used : Tubers.

Folk Use : Young tubers edible.

References : Krishna *et al.* (1999), Rao & Henery (1996), Watt (1889-1896).

Dioscorea triphylla (Linn.) Buch.-Ham. *ex* Wall.

Family : Dioscoreaceae.

Distribution : Konkan and Malabar Coast.

Part Used : Tubers.

Folk Use : Tubers bitter and eaten at the time of scarcity. They have to be well boiled to destroy the acridity.

Reference : Watt (1889-1896).

Remarks : Tubers acrid and be well boiled before consuming.

Dioscorea tuberosa Vell.

Family : Dioscoreaceae.

Distribution : N.W. Himalaya (Garhwal region).

Part Used : Tubers.

Folk Use : Tubers eaten after repeated boiling, washing and baking.

Reference : Gupta (1962).

Remarks : Tubers eaten after repeated washing and boiling to remove the acrid principle.

Dioscorea versicolor Wall.

Family : Dioscoreaceae.

Regional Names :

Ass. : Ghiti;
Mani. : Gheti;
Kum. : Ghangin, Ghenti.

Distribution : C. E. and W. Himalaya (Assam, Kumaon, Garhwal).

Reproductive Cycle : July-December.

Part Used : Tuberous Bulbs.

Folk Uses : Tubers - deliciously fragrant 'Yams' form a great part of food to the local tribals. To remove the acridity, they are always steeped for a night in water before cooking.

References : Bhargava (1959), Watt (1889-1896).

Remarks : Tubers acrid and unfit for use until steeped in water for a night before cooking.

Dioscorea villosa Linn.

Family : Dioscoreaceae.

Regional Name :

Mani. : Ha.

Distribution : E. Himalaya: 300-1,500 m.

Part Used : Tubers.

Folk Use : Tubers edible like other Yams.

Reference : Samant et al. (2001).

Dioscorea wallichii Hook. f.

Family : Dioscoreaceae.

Hindi and Regional Names:

Hindi : Man -Alu;
Beng. : Mou-Alu;
Ass. : Rui-Nihang;
Mani. : Ha;
Mal. : Neduvan.

Distribution : C. India and N.E. Himalaya.Kerala. Bihar. Orissa (wild, occasionally semicultivated).

Reproductive Cycle : October- February.

Part Used : Tubers.

Folk Uses : Tubers eaten raw, boiled or cooked. In Kerala, tubers consumed mainly by the 'Kani' tribe after repeated boiling so as to remove the acrid principle.

Active Constituents : Tubers comprise of carbohydrates, 18.9; fats, 0.1; moisture, 79.6 and proteins, 1.2g/ 100g.

References : Bennet *et al.* (1991), Das & Chatopadhyay (2001), Jain (1996), Radhakrishanan *et al.* (1996), Sinha (1996).

Remarks : Tubers acrid and must be repeatedly boiled before eating.

Dracocephalum heterophyllum Benth.

Family : Lamiaceae.

Regional Name :

Ladk. : Karram, Shanku, Zanda.

Distribution : Punjab. Trans-N.W. Himalaya (Ladakh) upto 3,500-4,500 m.

Part Used : Roots.

Folk Use : Roots used as vegetable by the tribals.

References : Ambasta (1986), Anonymous (1948-1976), Arora & Pandey (1996), Bennet *et al* (1991), Rana *et al.* (2005).

Dracontium polyphyllum Linn. A. Cunn. *ex* DC.

Family : Araceae.

Regional Names:

Tam. : Kaat- kurna;

Tel. : Adive Kanda.

Distribution : S.India.

Part Used : Tubers

Folk Use : Tubers edible.

References : Shortt (1877-1888), Watt (1889-1896).

Echinopsis nivensis Wall. *ex* Royle

Family : Asteraceae.

Distribution : C. Himalaya: 800-1,600 m.
PartUsed : Roots.
Folk Use : Roots edible.
Reference : Samant *et al.* (2001).

Eleocharis dulsis Trin.

Syn : *E. plantaginea* R.Br.; *E. tuberosa* Schult.
Family : Cyperaceae.

English and Hindi Names:

Eng. : Chinese Water Chestnut;
Hindi : Singapuri or Pani Chestnut.

Distribution : Entire Himalaya and throughout the greater parts of India.
Reproductive Cycle : October- December.
Part Used : Tubers.
Folk Uses : Tubers eaten raw as well as cooked and also served as salad and in soups. Its starch is comparable to that of the potato (*Solanum tuberosum*) and Cassava (*Manihot esculenta*).
Active Constituents : Root tubers yield albuminoids, 8.12; ash, 5.66; carbohydrates, 77.47; fats, 0.71; fibres, 3.70; and moisture, 4.34%.
References : Agarwal (2003), Ambasta (1986), Anonymous (1948-1976), Arora (1981), Arora & Pandey (1996), Chandra (1997).

Elephantopus scaber Linn.

Family : Asteraceae.

English, Hindi, Sanskrit and Regional Names:

Eng. : Elephant foot;
Hindi : Gobhi, Gojilata, Samudulan;
Sans. : Karipadam;
Beng. : Gojailata Shamudulum;
Mar. : Hastipata;
Tam. : Anashvoli;
Tel. : Hastikasala.

Distribution : Throughout the hotter parts of India.
Reproductive Cycle : September- January (Next year).
Part Used : Roots.
Folk Uses : Roots used as vegetable and also employed in the preparation of the local liquor 'Pachawi'.
References : Agarwal (2003), Anonymous (1948-1976), Chandra (1997).

Eria andamani Lindl.

Family : Orchidaceae.

Distribution : Andaman and Nicobar Islands.
Parts Used : Pseudobulbs. Rhizomes.
Folk Use : Both pseudobulbs and rhizomes edible.
Reference : Bhargava (1981).

Eriosema chinense Vog.

Family : Papilionaceae.

Hindi and Regional Names:
Hindi : Kodan;
Ass. : Soh - pan.

Distribution : E. and N.W. Himalaya (Garhwal to Assam) upto 2,000m, and in Orissa and Chotanagpur upto 1,500m.
Part Used : Tubers.
Folk Uses : Tuberous roots considered as tonic and eaten raw or cooked as vegetable.
References : Agarwal (2003), Ambasta (1986), Arora & Pandey (1996), Watt (1889-1896).

Eriosema himaliacum Ohaschi

Syn : *E. chinense* Sensu. Baker non Vog.; *E. tuberosum* (Buch.-Ham. *ex* D. Don) Wang. & Tang.

Family : Papilionaceae.

Regional Name :
Ass. : Sub -pen.

Distribution : N.E. Himalaya (Assam, Arunanchal Pradesh). M.P. upto 200 m.

Reproductive Cycle : August – November.
Part Used : Tender Roots.
Folk Use : Tender roots eaten raw or as vegetable.
References : Bennet *et al.* (1991), Samant *et al.* (2001).

Erycibe paniculata Roxb.

Family : Convolvulaceae.

Regional Names :

Ass. : Niguni-boallota;
Mal : Irimpiyatali;
Mani. : Uri;
Sant. : Kari;
Tam. : Unam-koddi;
Tel. : Puttapalatige.

Distribution : N.E. Himalaya and greater parts of India.
Part Used : Root Tubers.
Folk Use : Root tubers used as vegetable.
References : Jain (1996), Sinha (1996), Watt (1889-1896).

Erythropsis pallens Ridley

Syn : *Sterculia pallens* Wall.
Family : Sterculiaceae.

Regional Names:

Kum. : Khardala;
Lepa. : Kaphul.

Distribution : Throughout India.
Part Used : Roots.
Folk Use : Tender roots edible.
References : Anonymous (1948-1976), Arora & Pandey (1996).

Etlingera linguformis (Roxb.) Smith

Family : Zingiberaceae.

Distribution : N.E. Himalaya.
Part Used : Rhizomes.

Folk Uses	:	Rhizomes used as mouth freshner and also as food by the local tribes.
Reference	:	Tripathi & Goyal (2001).

Eulophia campestris Wall.

Family : Orchidaceae.

Hindi, Sanskrit and Regional Names:.

Hindi	:	Salibmisri;
Sans.	:	Amrita, Sudhamuli;
Beng.	:	Salbmisri, Sung - misri;
Guj.	:	Salum;
Mar.	:	Salibmisri;
Pub.	:	Salibmisri.

Distribution	:	N.W. Himalaya. Gangetic plains.
Reproductive Cycle	:	April-May.
Part Used	:	Tubers.
Folk Uses	:	Tubers of this and other allied species are used as salap, an easily digestable kind of farinaceous food which is highly nutritious and esteemed as a tonic.
References	:	Agarwal (2003), Arora & Pandey (1996), Watt (1889-1896).

Eulophia dabia (D.Don) Hochr.

Family : Orchidaceae.

Distribution	:	C. and N.W. Himalaya from Kumaon to Sikkim upto 2,000 m.
Part Used	:	Tubers.
Folk Use	:	Tubers edible and eaten cooked or raw.
Reference	:	Samant *et al.* (2001).

Euryale ferox Salisb.

Family : Nymphaeaceae.

English, Hindi and Regional Names:

Eng.	:	Foxnut, Gorgon-fruit;
Hindi	:	Makhana;
Beng.	:	Makhana;
Pub.	:	Jewar;
Tel.	:	Mellunipadma.

Distribution : Wetlands of the N.W. Himalaya and Peninsular India.

Part Used : Rhizomes.

Folk Use : Rhizomes used as vegetable.

References : Anonymous (1948-1976), Arora & Pandey (1996), Bennet et al. (1991).

Ferula asafoetida Linn.

Syn : *F.alliacea* Boiss.; *F. foetida* Regel.

Family : Apiaceae.

Hindi and Sanskrit Names:

Hindi : Hing.

Sans. : Balhika, Hingu.

Distribution : Kashmir.

Part Used : Latex or Gum-resin exuded from the living tap-root.

Folk Uses : Gum-resin employed to impart characteristic flavour in many food-stuffs particularly curries, sauces, pickles, etc. Essential oil, i.e. Asafoetida oil used as a fragrance component in perfumes.

Active Constituents : Components mainly responsible for the strong flavour are 2-butyl propenyl disulphide, 2-butyl-3-methyl-thioallyl disulphide,1-(methylthiopropyl)-1-propenyl disulphide, 1-(1-methyl thio-1-propenyl disulphide, di-2-butyl trisulphide, 2-butyl methyl trisulphide, di-2-butyl disulphide, di-2-butyl tetrasulphide, 2-butyl methyl disulphide and dimethyl trisulphide.

References : Ambasta (1986), Bhattacharjee (1998), Chevallier (1996), Greenish (1999), Jain (1968), Harborne & Baxter (2001), Mehtani & Ingle (1999), Rastogi & Mehrotra (1991,1993,1995), Singh & Pandey (1998), Watt (1889-1896).

Ferula galbaniflua Boiss.

Family : Apiaceae.

Hindi Name :

Hindi	:	Ganda -biroja.
Part Used	:	Latex or Gum-resin exuded from the living tap-root.
Folk Use	:	Used as a source of asafoetida.
References	:	Agarwal (2003), Watt (1889-1896).

Ferula narthrex Boiss.

Syn: *Narthrex asafetida* Falc.

Family : Apiaceae.

English , Hindi and Sanskrit Names:

Eng.	:	Asafoetida, Devil's Dung, Persian;
Hindi	:	Hing, Hingda, Hingra.
Sans.	:	Agudagandha, Balhika, Hingu, Ramatha.
Distribution	:	Native of Kashmir in India.
Part Used	:	Latex from tap-root.
Folk Uses	:	Another source of asafoetida used as a spice like 'sambhar' in S. India to give a scented smell to the food preparations.
Active Constituent	:	Organic sulphur compounds isolated.
References	:	Ambasta (1986), Anonymous (1999), Asolkar *et al.* (1992), Bhattacharjee (1998), Chopra *et al.* (1956), Kirtikar & Basu (1984), Raju (2000), Watt (1889-1896).

Flemingia strobilifera (Linn) R.Br.

Syn : *Moghania strobilifera* J.St. Hilaire.

Family : Papilionaceae.

Hindi and Regional Names:

Hindi	:	Kusrunt;
Ass.	:	Makhioti;
Bihar	:	Simbusk;
Kan.	:	Kumalu;
Mar.	:	Bundar, Kanphuti;
Tel.	:	Nallabaddu.

Distribution : N.W.Himalaya (Kumaon region): 800-2,000 m.

Part Used : Tender Roots.

Folk Use : Tender roots edible and used as vegetable by the locals.

References : Anonymous (1948-1976), Samant *et al.* (2001).

Flemingia tubersosa Dalz.

Family : Papilionaceae.

Regional Name :

Mar : Birmova.

Distribution : Grows mainly in Karnataka.

Part Used : Root Tubers.

Folk Uses : Astringent root tubers eaten raw or roasted.

Active Constituents : Root tubers reveal the occurrence of albuminoids, 13.04; ash, 3.44; carbohydrates (starch), 40.12; fibres, 12.16; soluble yellow resin, 1.64; sugar and gum, 25.47 % and traces of tannins.

References : Agarwal (2003), Anonymous (1948-1976).

Flemingia vestita Benth. *ex* Baker

Syn: *F. procumbens* Roxb.; *Moghania procumbens* (Roxb.)Mukerjee; *Moghania tubersosa* (Dalz.) O. Kuntze; *Moghania vestita* O. Kuntze

Family : Papilionaceae.

Hindi and Regional Name :

Hindi : Sophlang ;
Khasi : Sophlang.

Distribution : N.W. Himalaya, ascending to 2,900m.

Reproductive Cycle : May-November.

Part Used : Tubers.

Folk Uses : Tender root tubers used as vegetable or eaten raw by N.E. tribals. Outer skin of tubers has an agreeable flavour.

Active Constituents : Tubers rich in genistein, daidzein, formononetin and pseudobaptigenin.

References : Agarwal (2003), Ambasta (1986), Arora (1981), Arora & Pandey (1996), Bennet (1987), Negi & Pant (1992), Samant *et al.* (2001), Watt (1889-1896).

Gagea persica Boiss.

Family : Liliaceae.

Distribution : Entire Himalaya.

Part Used : Bulbs.

Folk Use : Bulbs edible.

Reference : Arora & Pandey (1996).

Gustrochilus pandurata Ridley

Family : Zingiberaceae.

Distribution : Andamans. Konkan.

Part Used : Bulbs.

Folk Use : Bulbs used as condiments.

References : Mittre (1981), Kirtikar & Basu (1984).

Geranium wallichianum Don. *ex* Sweet

Family : Geraniaceae.

English and Regional Names:

Eng. : Wallich Cranesbill;
Kash. : Kao- ashud;
Pub. : Laljhari, Lilizhari.

Distribution : N.W. Himalaya (Kumaon & Garhwal): 2,000-4,000 m.

Part Used : Roots.

Folk Use : Roots edible.

Reference : Samant *et al.* (2001).

Glycyrrhiza glabra Linn.

Family : Papilionaceae.

English , Hindi, Sanskrit and Regional Names:

Eng. : Licorice, Liquorice;
Hindi : Jethi-Madh, Mulhatti;

Sans. : Jalayashti, Madhuka, Yashti-Madhu;
Beng. : Jaishbomadhu, Jashtimadhu;
Guj. : Jethi-madha;
Mar : Jeshta madha;
Mal. : Iratimaduram;
Tam. : Atimaduram;
Tel. : Atimadhuram, Yashtimadukan.

Distribution : Kashmir Himalaya. Cultivated especially in sub-tropical and tropical regions.

Parts Used : Rhizomes. Roots.

Folk Uses : Rhizomes, roots and liquorice extract mainly used as a sweetening agent for flavouring cigarettes, snuff, tobacco, confectionery, beverages, jams, marmalades, sweet sauces, reconstituted vegetable proteins, chocolates, chewing gum, diet colas and beer.

Active Constituents : Characteristic aroma of the volatile oil of liquorice is due to estragole, anethole, eugenol, indole, ã-nonalactone and cumin alcohol.

References : Ambasta (1986), Anonymous (1999), Bhattacharjee (1998), Chevallier (1996), Chopra *et al.* (1968), Greenish (1999), Kirtikar & Basu (1984), Mehtani & Ingle (1999), Rastogi & Mehrotra (1991, 1993, 1995), Sharma & Sood (1997), Sinha (1996), Watt (1889-1896).

Gonatanthus pumilus (D.Don) Engl.

Syn : *G. sarmentosa* Klotzsch.& Otic.

Family : Araceae.

Hindi Name :

Hindi : Jangli Pindalu.

Distribution : Arunanchal Pradesh: 1,500-2,200 m.

Part Used : Tuberous Roots.

Folk Use : Root tubers eaten as vegetable after prolonged boiling.

References : Arora & Pandey (1996), Rana *et al.* (2005), Samant *et al.* (2001).

Habenaria clavigera (Lindl.) Dandy

Syn : *H. densa* Lindl.

Family : Orchidaceae.

Distribution : Temperate Himalaya.

Part Used : Tubers.

Folk Use : Tubers edible.

Reference : Arora & Pandey (1996).

Habenaria commelinifolia Wall. *ex* Lindl.

Family : Orchidaceae.

Hindi Name :

Hindi : Choora.

Distribution : N.W. Himalaya and almost throughout India from Punjab eastwards to Bihar, Orissa and S. India.

Part Used : Tubers.

Folk Uses : Tubers edible and furnish a good quality salap.

References : Agarwal (2003), Anonymous (1948-1976), Kirtikar & Basu (1984).

Habenaria susannae (L.) R. Br.

Syn : *Platanthera susannae* (L.) Lindl.

Family : Orchidaceae.

Hindi and Regional Names :

Hindi : Hukakanda;
Bomb : Waghchura;
Nagp. : Hukakanda.

Distribution : Throughout India: 1,500- 2,000 m.

Part Used : Tubers.

Folk Use : Large-sized tubers edible

References : Agarwal (2003), Anonymous (1948-1976), Bajpayee & Dixit (1996).

Remarks : This plant is becoming extinct rapidly due to excessive digging of the tubers for edible purposes by the tribals.

Hedychium coronarium Koenig.

Syn: *H. flavum* Roxb.

Family : Zingiberaceae.

English and Hindi and Names:

Eng. : Common Ginger Lily, Garland Flower, Yellow Ginger Lily;

Hindi : Karchura;

Mani. : Takhelu.

Distribution : Throughout moist parts of India upto 2,400m. Also grown in gardens.

Reproductive Cycle : June-November.

Part Used : Rhizomes.

Folk Uses : Rhizomes yield edible arrowroot like starch and also used as tonic. Rhizome and the young shoots of the plant are an essential ingredient of the "Seasone Yongchak" (Yongchak is a Manipuri name of special food prepared from these material alongwith other spices) and 'Yongchak-Iromba' (a special food of Assam and Manipur tribes prepared from many spices and fish alongwith 'Yongchok').

Active Constituents : Analysis of rhizome reveals albuminoids, 1.65 ; ash, 2.3 ; cellulose, 29.6; extracted matter,0.91; fats, 0.33 ; gum, 13.75; organic acid 5.5; resinous acid, 3.66; starch, 30.02%, which is similar to the arrowroot starch. Organic acids and alcohol present are linalool, methyl benzoate, cis-jasmone, eugenol, (E) isoeugenol .

References : Agarwal (2003), Ambasta (1986), Anonymous (1948-1976), Arora & Pandey (1996), Aswal (2001), Asolkar *et al.* (1992), Bennet *et al.* (1991), Chandra (1997), Singh *et al.* (2001b), Jain (1996), Rastogi & Mehrotra (1990, 1995), Sinha (1996).

Hedychium spicatum Buch.-Ham. *ex* Smith (Pl. 7A)

Family : Zingiberaceae.

English, Hindi, Sanskrit and Regional Names:

Eng.	:	Spiked Ginger Lily, The Lesser Galangal of Ainslie;
Hindi	:	Gandhapalashi, Kapurakachari, Sitruti;
Sans.	:	Amlaharidra, Durva, Gandha, Karpurakachali, Himaja, Palashi;
Beng.	:	Kapura kachli;
Mani.	:	Lamtak-helli;
Guj.	:	Kapura kachli ;
Mar	:	Kapura kachli ;
Pub.	:	Sheduri ,Sitruti;
Tam.	:	Kishngu, Shimai- kich- chilli.

Distribution	:	C. , W. and subtropical Himalaya upto 2,000 m.
Reproductive Cycle	:	May-September.
Part Used	:	Rhizomes.
Folk Uses	:	Aromatic rhizomes considered appetizer and tonic, and used in the special preparation of 'Yongchok- Iromba' in Manipur.
Active Constituents	:	Rhizomes contain hedychenone, diterpene-6-oxo-labda-7,11,13-trien-16-oic acid lactone (I).
References	:	Agarwal (2003), Ambasta (1986), Anonymous (1948-1976), Arora & Pandey (1996), Aswal (2001), Kirtikar & Basu (1984), Rastogi & Mehrotra (1993,1995), Sarin (1989), Singh (1999), Singh *et al.* (2001b), Watt (1889-1896).

Helianthus tuberosus Linn.

Family : Asteraceae.

English, Hindi and Regional Names:

Eng.	:	Girasole,Jerusalam-artichoke, Topinamber;
Hindi	:	Artichuk, Hathchuk;
Beng.	:	Brahmokha;
Pub.	:	Hathipicha.

Distribution : N.E. and N.W. Himalaya from Garhwal to Assam, W. India and Bengal.

Part Used : Tuberous Rhizomes.

Folk Use : Tubers eaten raw or boiled, also pickled and made into chips or grounded into flour which is considered equivalent to potato in food values.

Active Constituents : Tuberous rhizome comprises of ash, 0.85-2.5; carbohydrates, 8.0-18.0; fats, 0.11-0.44; fibres, 0.3-3.0; nitrogen free extract, 13.8-18.8; proteins, 0.9-3.25; minerals, (Ca, 0.023; P, 0.099%; Fe, 3.4mg/100g,); Al, Cl, I, K, Mg, Na, Zn, (12mg /100g); moisture, 71.6-84.2 and traces of vitamins A, B, and C.

References : Agarwal (2003), Ambasta (1986), Anonymous (1948-1976), Bajpayee & Dixit (1996), Rana *et al.* (2005) Singh *et al.* (1996).

Hemidesmus indicus (L.) Schult.

Family : Asclepiadaceae.

English, Hindi, Sanskrit and Regional Names:

Eng. : Indian Sarsaparilla;
Hindi : Anantmul, Sala Magrabu;
Sans. : Anantmula;
Beng. : Anantmul;
Guj. : Durid, Sariva, Upalas;
Tam. : Nannari;
Tel. : Sungadali-pala.

Distribution : Almost throughout India upto 1,000 m.

Reproductive Cycle : August-December.

Part Used : Roots.

Folk Use : A syrup made from roots used as a flavouring agent in the preparation of 'Sherbat' having cooling property.

References : Anonymous (1948-1976), Bennet *et al.* (1991).

Heracleum pinda Dalz. & Gibs.

Family : Apiaceae.

Regional Name:

Mar. : Pundha.

Distribution : Goa. Maharashtra (Chhota Nagpur).

Part Used : Roots.

Folk Use : Roots edible.

References : Arora & Pandey (1996), Vartak (1981).

Houttuynia cordata Thunb.

Family : Saururaceae.

Regional Names :

Ass. : Machandari;

Megh. : Jamirdoh (Khasia and Jantia Tribe).

Distribution : N.E and sub tropical Himalaya upto 1,800 m.

Part Used : Rhizomes.

Folk Uses : Rhizomes eaten raw, boiled or roasted alongwith the shoots and leaves, also these are used in curries or cooked with rice, sweet potato or beans.

References : Agarwal (2003), Ambasta (1986), Anonymous (1948-1976), Arora (1981), Arora & Pandey (1996), Bennet *et al.* (1991), Borthakur (1996), Jain (1996), Joseph & Kharkongar (1996).

Hypoxis aurea Lour.

Family : Amaryllidaceae.

Hindi Name :

Hindi : Kinda.

Distribution : Himalaya and other hilly tracts of India: 1,000-2,000 m.

Part Used : Rootstocks.

Folk Uses : Rootstocks edible and also considered as tonic.

References : Agarwal (2003), Ambasta (1986), Samant *et al.* (2001).

Ipomoea aquatica Forsk.

Syn : *Convolvulus reptans* Roxb.; *I. reptans* Poir.
Family : Convolvulaceae.

English, Hindi, Sanskrit and Regional Names:

Eng. : Swamp Cabbage;
Hindi : Kalim-sag ;
Beng. : Kalmi- sag ;
Guj. : Nalani Bhaji ;
Mar : Nadishaka, Nalichi Bhaji;
Tam. : Sarkivalli;
Tel. : Tuti-kara.

Distribution : An aquatic herb occurring throughout India.

Reproductive Cycle : Throughout the year.

Part Used : Rhizomes. Roots.

Folk Uses : Roots and rhizomes universally eaten as vegetable. In Assam rhizomes and roots cut into pieces are boiled in acid alongwith fish for the preparation of 'Singju'- a special salad of the 'Meitei' community and 'Iromba' - a special delicacy in major meals prepared from a number of plant ingredients boiled alongwith the fruits of Capsicum annuum (chilli) and 'Nagari' ('Nagari' refers to the fermented preparation of the following three species of fish: *Punitius sophore* Ham.; *P.chola* Ham; *P. ticto* Ham.) so as to enhances the palatability of the dish.

References : Agarwal (2003), Ambasta (1986), Anonymous (1948-1976), Arora (1981), Arora & Pandey (1996), Bennet *et al.* (1991), Gammie (1902), Gupta & Kanodia (1968), Rana (2004), Raju (2000), Saini (1996), Saxena (1979), Singh *et al.* (2001b), Watt (1889-1896).

Ipomoea batatus Lamk. (Pl. 7B)

Syn : *Batatus edulis* Chois.; *Convolvulus batatus* Linn.; *C. edulis* Thunb; *C. esculantus* Salisb.

Family : Convolvulaceae.

English, Hindi, and Regional Names:

Eng.	:	Sweet-potato;
Hindi	:	Mitha Alu, Shakarkandi;
Beng.	:	Chine Alu, Rangalu;
Guj.	:	Shakaria;
Mal.	:	Kappa-kelanga;
Kan.	:	Ganasu.

Distribution : Cultivated all over India.

Part Used : Tubers.

Folk Uses : Sweet tubers eaten by all class of the natives either in curries or after simply roasting. It is also consumed raw or made into pudding in the usual traditional style with sugar, egg and milk.

Active Constituents : Root tubers comprise of ash, 0.6-1.5; carbohydrates, 31.0; fats, 0.3; fibres, 0.6-2.5; minerals, 1.0%, (Ca,12; P,49; Fe, 40mg /100g) moisture, 68.5 and proteins 1.2 %.

References : Agarwal (2003), Ambasta (1986), Anonymous (1948-1976), Watt (1889-1896).

Ipomoea cairica (Linn.) Sweet

Syn : *I. palmata* Forsk.

Family : Convolvulaceae.

English and Hindi Names:

Eng.	:	Railway creeper;
Hindi	:	Rellata.

Distribution : Almost throughout India.

Part Used : Tuberous Roots.

Folk Uses : Tuberous roots edible and also used in the preparation of local wine by the tribals.

References : Agarwal (2003), Anonymous (1948-1976).

Remarks : Roots considered carcinogenic but still used by the tribals.

Ipomoea digitata Linn.

Family : Convolvulaceae.

Hindi, Sanskrit, and Regional Names:

Hindi : Bilaikand;
Sans. : Bhumikushmunda;
Beng. : Bhumikumara;
Mar : Bhukobola;
Tam. : Palmudangi;
Tel. : Bhuchakaragadda.

Distribution : E. Himalaya. Coastal India. Karnataka.

Part Used : Rhizomes.

Folk Uses : Rhizomes used as vegetable and considered tonic.

Active Constituents : Rich in vit. A (carotene), 6mg/100g.

Reference : Agarwal (2003).

Ipomoea muricata Linn.

Syn : *Convolvulus muricatus* Linn.; *Calonyction muricatum* G.Don.

Family : Convolvulaceae.

Regional Name :

Eng. : Wallich Cranesbill;
Bomb. : Gariya.

Distribution : Common throughout India; semicultivated as well as wild.

Part Used : Roots.

Folk Use : Roots used as famine food.

Reference : Watt (1889-1896).

Ipomoea paniculata R.Br.

Family : Convolvulaceae.

Hindi Name :

Hindi : Bhumikushmund.

Distribution : Grows throughout the moist parts of India.
Part Used : Roots.
Folk Use : Roots edible and also used as sweet tonic.
Reference : Agarwal (2003).

Ipomoea tuba Schl.

Syn : *I. grandifolia* Lamk.
Family : Convolvulaceae.

Hindi Name :

Hindi : Sundrilata.

Distribution : Coastal areas of India.
Part Used : Rhizomes.
Folk Use : Rhizomes edible.
Reference : Agarwal (2003).

Isoetes coromandeliana Linn.

Family : Isoetaceae.

Regional Name :

Mani. : Sorban.

Distribution : N.E.Himalaya (Assam and Manipur).
Part Used : Rhizomes.
Folk Uses : Rhizomes alongwith the aerial plant parts are used in the preparation of the 'Iromba' and 'Shingju'.
Reference : Singh & Viswanathan (1996), Singh *et al.* (2001b).

Kaempferia angustifolia Rosc.

Family : Zingiberaceae.

Hindi Names :

Hindi : Kanjan-Bura, Mudu-nirbisha.

Distribution : A native of Bengal.
Part Used : Roots.
Folk Use : Roots used as food at the time of scarcity.
References : Mehra *et al.* (1975), Watt (1889-1896).

Kaempferia galanga Linn.

Syn: *Alpinia sessilis* Kon.
Family : Zingiberaceae.

English , Hindi, Sanskrit and Regional Names:

Eng. : Black Thorn;
Hindi : Chandramula;
Sans. : Chandramulika, Sugandhavacha;
Beng. : Chandermula, Humula;
Tam. : Kotsjolam;
Tel. : Kachoram.

Distribution : Throughout the hotter parts of India often cultivated in gardens.

Part Used : Rhizomes.

Folk Uses : Rhizomes camphoraceous odoured and used as condiment and spices for flavouring rice and other food items. These are also used as a ingredient of 'Pan' (betel) as masticatory and flavouring material.

Active Constituents : These are deoxypodophyllotoxin, ethyl p-methoxy-trans-cinnamete, monoterpene ketone-car-3-en-5-one (rhizome) and Et-p-MeO-trans-cinnamate (roots).

References : Ambasta (1986), Anonymous (1948-1976), Asolkar et al. (1992), Islam (1996), Rastogi & Mehrotra (1990), Tripathi & Goyal (2001), Watt (1889-1896).

Kaempferia rotunda Linn.

Family : Zingiberaceae.

Hindi, Sanskrit and Regional Names:

Hindi : Bhuichampa;
Sans. : Bhuchampaca, Bhumichampa;
Beng. : Bhuchampa;
Guj. : Bhuchampo;
Mar : Bhuchampa;
Tam. : Konda Kalava;
Tel. : Bhucham - pakamu.

Distribution : Throughout India mainly Assam, Meghalaya, Mizoram and in Tripura.

Reproductive Cycle : April-June.
Part Used : Rhizomes.
Folk Use : Tuberous rhizomes used as a flavouring material.
Active Constituents : Flavouring compounds, cineole and methyl chavicol isolated from rhizome.
References : Ambasta (1986), Chandra (1997), Kirtikar & Basu (1984), Mehtani & Ingle (1999), Watt (1889-1896).

Kedrostis rostrata (Rottl.) Cogn.

Syn : *Rhynchocarpa foetida* C.B.Clarke.
Family : Cucurbitaceae.

Hindi and Regional Names:

Hindi : Nuruk-bel;
Tam. : Kuku madunda;
Tel. : Appakoray.

Distribution : Endemic to Deccan Peninsula.
Part Used : Roots.
Folk Use : Sweet mucilaginous roots eaten at the time of scarcity.
Reference : Agarwal (2003).

Lasia spinosa (Linn.) Thaw., D.C.

Syn : *Draconetum spinosum* Lin.; *Lasia heterophylla* Linn.; *Lasia macrophylla* Schott.; *Pothos lasia* Roxb.; *Pothos spinosus* Ham.
Family : Araceae.

Regional Names :

Beng. : Kanta -katachu ;
Mun. : Janum-saru;
Santh : Kanta -saru;
Tel. : Mulaseri.

Distribution : N.E. Himalaya and subhimalayan regions. W. Bengal.
Part Used : Tuberous Rhizomes.
Folk Uses : Small tubers eaten raw or after boiling and

thorough washing. Roots also used as vegetable and in curries.

References : Arora (1981), Arora & Pandey (1996), Bennet *et al.* (1991), Borthakur (1996), Rao & Henery (1996), Watt (1889-1896).

Leea edgeworthii Santapu

Family : Vitaceae.

Hindi Name :

Hindi : Kumala , Kumali.

Distribution : Entire Himalaya and throughout India.

Part Used : Tuberous Roots.

Folk Use : Tuberous roots consumed after boiling.

Reference : Arora & Pandey (1996).

Leptadenia pyrotechnica (Forsk.) Decne.

Syn : *Leptadenia spartium* Wt. Khip.

Family : Asclepiadaceae.

Hindi and Regional Names:

Hindi : Klp;
Guj. : Khip;
Pub. : Kip;
Tam. : Pallakudy.

Distribution : Sandy soils of N. and W. India. Also in coastal areas.

Part Used : Tuberous Roots.

Folk Use : Tuberous roots consumed as vegetable.

References : Agarwal (2003), Ambasta (1986), Arora & Pandey (1996).

Leptadenia raticulata Wight & Arn.

Family : Asclepiadaceae.

Hindi Name :

Hindi : Dori.

Distribution : Subhimalayan tract of Punjab and U.P., and in Deccan Peninsula.

Part Used : Tuberous Roots.

Folk Uses : Tender roots alongwith its leaves and shoots eaten as vegetable. Also these are used in preparing bread by mixing its boiled roots with flour of 'bajri' (millet) or 'jowari' (*Sorghum vulgare* Pers.).

References : Agarwal (2003), Gammie (1902), Shortt (1887-1888), Watt (1889-1896).

Leucojum aestivum Linn.

Family : Amaryllidaceae.

Distribution : Grown as an ornamental plant in S. India

Part Used : Bulbs.

Folk Use : Bulbs eaten after cooking.

Reference : Ambasta (1986).

Remarks : Bulbs considered emnetic as it contains sistrin-like carbohydrates.

Lilium nepalensis D. Don.

Family : Liliaceae.

Distribution : C. Himalaya.

Part Used : Starchy Bulbs.

Folk Use : Starchy bulbs edible and used as vegetable.

Reference : Arora & Pandey (1996).

Limnanthemum indicum (Linn.) Thw.

Syn : *Nymphoides indicus* Roxb., Kuntze.

Family : Nymphaeaceae.

Hindi Name :

Hindi : Barachuli.

Distribution : Grown throughout India in ponds.

Parts Used : Rhizomes. Roots.

Folk Use : Rhizomes and roots used as vegetable.

References : Agarwal (2003), Arora & Pandey (1996).

Litsea glutinosa (Lour.) C.B.Robins.

Syn : *L. chinesis* Lam.; *L. sebifera* Pers.

Family : Lauraceae.

Hindi and Regional Names:

Hindi	:	Garbijaur, Maida Lakri ;
Beng.	:	Garur,Kukur chita , Ratun;
Guj.	:	Maida, Lakdi;
Mar	:	Maida, Lakdi;
Pub.	:	Chandra, Maida lakri, Maida sak;
Tel.	:	Nara- alagi.

Distribution : Throughout India.

Part Used : Roots.

Folk Use : Roots used as tonic and vegetable.

Reference : Agarwal (2003).

Manihot esculenta Crantz. (Pl. 7A)

Syn : *Janipha manijhot* Kunth.; *Manihot aipi* Pohal.; *M. dulsis* Roxb.; *M.palmate* Muell.; *M. utillisima* Pohal.

Family : Euphorbiaceae.

English, Hindi and Regional Names:

Eng.	:	Cassava, Manioc, Tapioca ;
Hindi	:	Alu-Kand, Shimla -Alu ;
Ass.	:	Simul -Alu ;
Mal.	:	Marachini;
Kan.	:	Marragenasu;
Tam.	:	Marravalli;
Tel.	:	Karrapendalamu.

Distribution : Cultivated in warmer and moist regions of India: 300-1,200 m.

Part Used : Tuberous Roots.

Folk Uses : Roots edible and used as vegetable. Also these yield true 'Tapioca' and 'Cassava' for which tubers are made into pulp after proper washing, filtration and drying, and starchy flour is made and packed as 'Cassava'. 'Tapioca' meal is prepared from squeezed out juice dried in air without being roasted. Rice is also made as

'Tapioca Manioc' for which the rind is cut into small pieces and washed in running water for one or two days to detoxify it and followed by drying of the pieces either in smoke or sun. Before cooking in a sauce of palm oil the pieces are carefully soaked in water to give them a spongy consistency similar to mushrooms. The rind of 'Manioc' is also eaten as such during famine.

Active Constituents : Fresh tubers reveal carbohydrates, 38.7; fats, 0.2 ; fibres, 0.6-2.5 ; minerals, 1.0 %, (Ca, 10.64 ; P,15.58 ; Fe, 0.66; Mg, 7.35; S, 3.73; Si, 0.94; Cl, 2.75 mg/100g, and Iodine in traces); moisture, 69.4 ; proteins 0.7 %; riboflavin, 0.01; thiamine, 0.045 and nicotinic acid, 0.3mg / 100g. Starch content varies from 77.5- 85.0 % of the total carbohydrates.

References : Agarwal (2003), Anonymous (1948-1976), Bajpayee & Dixit (1996), Samant *et al.* (2001), Singh *et al.* (1999), Watt (1889-1896).

Manihot glaziovii Muell.

Family : Euphorbiaceae.

English Name :

Eng. : Ceara Rubber, Manihot Rubber.

Distribution : N.E.Himalaya (Assam). S. India.

Part Used : Tuberous Roots.

Folk Use : Tuberous roots used as vegetable.

Reference : Agarwal (2003).

Maranta arundinacea Linn.

Syn : *Maranta indica* Jussac.; *M. ramosissima* Wall.

Family : Marantaceae.

English, Hindi and Regional Names:

Eng. : Arrowmehl, W. Indian Arrowroot;

Hindi : TIkhur;

Beng. : Aravut;
Guj. : Aravut;
Mar : Tavkil;
Kan. : Tavaksha;
Tam. : Kuamou.

Distribution : Wild in some parts of India but cultivated in Assam, Bengal, Bihar, Kerala, Orrisa, U.P.

Part Used : Rhizomes.

Folk Uses : Rhizomes eaten boiled or roasted, and also prescribed as food for invalids. Fine flour of the rhizomes called arrowroot is employed in the preparation of biscuits, cakes, pudding, pastries and jellies. For the preparation of arrowroot, rhizomes after washing are pounded to pulp in a wooden mortar. The pulp is then thrown into a large vessel of water which becomes turbid and milky. It is filtered into another vessel separated from the fibrous mass. The milky water is allowed to sediment, resulting in the fine 'arrowroot' starch. It is dried into fine sheets of the paper by exposing to the sun. It forms a good jelly, and considered as a good tonic and thickener for other nutritious substances.

Active Constituents : Fresh rhizomes yield ash, 0.9; carbohydrates (starch), 27.8; dextrin and sugars, 2.1; fats, 0.2 ; fibres,3.9; moisture, 63.4 and proteins 1.6%.

References : Agarwal (2003), Ambasta (1986), Anonymous (1948-1976), Watt (1889-1896).

Maranta dichotoma Wall.

Family : Marantaceae.

Hindi Name :

Hindi : Desi Arrowroot.

Distribution : Cultivated in the warmer parts of India and Coromandal Coast.

Part Used : Rhizomes.
Folk Use : Rhizomes edible and used for making confectionery items.
References : Agarwal (2003), Watt (1889-1896).

Marrubium roylei (Wight.)

Family : Lamiaceae.

Hindi Name :

Hindi : Murkula.

Distribution : Entire Himalaya.
Part Used : Roots.
Folk Use : Roots eaten by 'Lepachas' (N.E. tribals).
Reference : Ambasta (1986).

Marsdenia roylei Wt.

Family : Asclepiadaceae.

Regional Names :

Kum. : Murkula;
Garh. : Shengori.

Distribution : N.E. (Assam) and temperate Himalaya.
Reproductive Cycle : May- July.
Part Used : Roots.
Folk Uses : Roots eaten by the 'Lepachas' (N.E. tribals).
References : Agarwal (2003), Ambasta (1986), Anonymous (1948-1976), Chandra (1997), Arora & Pandey (1996).

Marsilea quadrifolia Linn.

Family : Marsileaceae.

Regional Names :

Beng. : Surni-Shak;
Pub. : Godhi , Tripatra.

Distribution : Banks of the rivers and ponds of lower Himalaya, and from Punjab to Bengal.
Part Used : Rhizomes.

Folk Use : It is regularly eaten as a pot herb by the natives in Punjab and other parts of India.

References : Anonymous (1948-1976),Watt (1889-1896).

Meconopsis aculeata Royle

Family : Papaveraceae.

Distribution : Trans-N.W .Himalaya (J.&K. & H.P.): 3,500-4,800 m.

Part Used : Roots.

Folk Uses : Roots eaten raw or used as vegetable by the local tribals.

Reference : Samant *et al.* (2001).

Meconopsis paninculata D.Don

Syn : *M. nepalensis* Hook. f. & Thoms. non DC.

Family : Papaveraceae.

Distribution : Nepal to Sikkim Himalaya upto 3,500 m.

Part Used : Roots.

Folk Uses : Roots regarded as narcotic, and eaten by the local herdsman's after peeling the outer rind.

References : Agarwal (2003), Anonymous (1948-1976),

Melothria heterophylla (Lour.) Cogn.

Syn : *Zehneria umbellata* Thw.

Family : Cucurbitaceae.

Hindi and Regional Names:

Hindi : Anantmul, Tarali;
Beng. : Kudari;
Khasi : Soh-Khia-Phlang;
Mani. : Lamgi-Shebot;
Pub. : Bankakri;
Tam. : Pullivanji;
Tel. : Thia-Donda.

Distribution : N.E. Himalaya (Manipur) and Hilly parts of India.

Part Used : Roots.

Folk Uses : Roots edible.
References : Agarwal (2003), Anonymous (1948-1976), Rana *et al.* (2005), Sinha (1996).

Merremia mammosa (Lour.) Hallier.f.

Syn : *Ipomoea gomezii* C.B.Clarke.
Family : Convolvulaceae.

Distribution : N.E. Himalaya (Assam). Andaman and adjacent areas.
Part Used : Tubers.
Folk Use : Tubers edible.
References : Arora & Pandey (1996), Anonymous (1948-1976).

Merremia rhyncorhiza (Dalz.) Hallier.f.

Syn : *Ipomoea rhyncorhiza* Dalz.
Family : Convolvulaceae.

Distribution : W. Ghats.
Part Used : Tubers.
Folk Use : Tubers edible after boiling.
References : Arora & Pandey (1996), Anonymous (1948-1976).

Merremia tubersosa (Linn.) Rendle

Syn : Ipomaea tubersosa Linn.
Family : Convolvulaceae.

English and Hindi Names:
Eng. : Brazilian Jalap ;
Hindi : Prasirini.

Distribution : Grown in Indian gardens as an ornamental plant.
Part Used : Tuberous Roots.
Folk Use : Tuberous roots eaten.
References : Agarwal (2003), Ambasta (1986), Anonymous (1948-1976), Arora & Pandey (1996).

Mirabilis jalapa Linn.

Family : Nyctaginaceae.

English and Regional Names:

Eng. : Four 'O' Clock plant, Marval of Peru;
Guj. : Gubbaji;
Kan. : Sanja mallige ;
Tam. : Andhimalligai ;
Tel. : Chandra lata.

Distribution : Trans-N.W. Himalaya (J.&K. & H.P.): 1,500-2,500 m. Also grown as an ornamental plant in Indian gardens.

Part Used : Roots.

Folk Use : Roots edible and eaten raw.

Active Constituents : Glucose and arabinose are the chief constituents of roots.

References : Rao & Henery (1996), Samant *et al.* (2001).

Momordica dioica Roxb.

Family : Cucurbitaceae.

Hindi and Regional Names:

Hindi : Golkandra, Jangli Kla, Kaksa ;
Guj. : Kartoli.

Distribution : Almost throughout India and in Himalaya upto 1,500 m.

Reproductive Cycle : June-August.

Part Used : Roots.

Folk Use : Roots edible and eaten as vegetable after boiling and cooking.

References : Anonymous (1948-1976), Arora & Pandey (1996), Bennet *et al.* (1991), Vartak (1981).

Moringa oleifera Lam.

Syn : *Moringa pterygosperma* Gart.

Family : Moringaceae.

English, Hindi and Sanskrit Names:

Eng. : Drum-stick Tree, Horse Radish Tree;
Hindi : Asainjana, Mungna, Shajna;
Sans. : Shobhanjana.

Distribution : Subhimalayan tract. Semicultivated in plains of India.

Part Used : Roots.

Folk Use : Roots used as a condiment.

References : Anonymous (1999), Arora & Pandey (1996).

Mucuna atropurpurea DC.

Family : Papilionaceae.

Distribution : E., C. and N.E. Himalaya upto 1,200 m.

Part Used : Roots.

Folk Use : Roots edible.

Reference : Samant *et al.* (2001).

Nelumbo nucifera Gartn.

Syn : *Cyamus nelumbus* Smith.; *Nelumbium speciosum* Willd.; *Nelumbo indica* Poir.; *Nymphaea asiaticum* Rich.; *Nymphaea nelumbo* Linn.

Family : Nymphaeaceae.

English, Hindi, Sanskrit and Regional Names:

Eng. : Lotus, The Sacred Lotus;
Hindi : Kamal, Kanval, Kanwal;
Sans. : Kamal, Nilofer, Rajiv, Padam, Pushkar;
Ass. : Uttpala -Padam;
Beng. : Padama;
Guj. : Suriya Kamal;
Mal : Thamara;
Mani. : Ikai-Thabi, Thambal;
Kan. : Tavare-gadde;
Tel. : Erra-tamara-veru, Kalung -erra- tamara.

Distribution : Wetlands of Himalaya and all over India.

Reproductive Cycle : June –September.

Part Used : Rhizomes.

Folk Uses : Rhizomes cooked or pickled and also yield nutritious arrowroot. These are sold in markets as a famine food. The fresh rhizome also eaten after roasting while its dried slices are used in curry or fried as chips as well as preserved in frozen condition and used as an ingredient in precooked food.

Active Constituents : Analysis of fresh rhizomes reveals ash, 0.6-1.5; carbohydrates (starch, 9.25; sucrose, 0.41; reducing sugars, 1.56); fats, 0.8; fibres, 0.6-2.5; minerals, 1.0 (Ca, 0.06; P and Fe in traces); water ,83.80 and proteins 2.70 %.

References : Agarwal (2003), Ambasta (1986), Anonymous (1948-1976), Arora (1981), Arora & Pandey (1996), Bennet *et al.* (1991), Borthakur (1996), Das *et al.* (1996), Krishna *et al.* (1999), Raju (2000), Rana *et al.* (2005), Rao & Henery (1996), Sahu (1996), Samant *et al.* (2001), Singh *et al.* (2001b), Watt (1889-1896).

Nelumbo speciosum Willd.

Family : Nymphaeaceae.

English and Regional Names :

Eng. : Pink Water Lily;
Ass. : Tharo.

Distribution : Wetlands of N. E. Himalaya.

Part Used : Rhizomes.

Folk Uses : Rhizomes eaten as vegetable. These alongwith its petiole are used in the preparation of 'Shingju'- a special salad of the 'Meitei' community. To enhance the taste, powdered seeds of *Anisomeles indica* are added alongwith a little roasted flour of gram.

References : Gammie (1902), Singh *et al.* (2001b), Watt (1889-1896).

Nephrolepis auriculata (Linn.) Triman.

Family : Nephrolepidaceae.

Distribution : N.W. Himalaya from Kumaon to Garhwal upto 2,000 m.

Part Used : Tubers.
Folk Use : Tubers edible and used as vegetable by the tribals.
Reference : Samant *et al.* (2001).

Nephrolepis biserrata Schott.

Syn : *N. acuta* Presl.
Family : Nephrolepidaceae.

Hindi Name :

Hindi : Parni.

Distribution : N.W. Himalaya: 1,000-2,000m. Maharastra. S. India upto 1,500 m.
Part Used : Rhizomes.
Folk Use : Rhizomes edible.
References : Agarwal (2003), Ambasta (1986), Samant *et al.* (2001).

Nephrolepis cordifolia (Linn.) Kunth.

Family : Nephrolepidaceae.

Distribution : N. E. Himalaya.
Part Used : Rhizomes.
Folk Use : Rhizomes edible and used as vegetable by the tribals.
Reference : Singh & Viswanathan (1996).

Nymphaea alba Linn.

Family : Nymphaeaceae.

English and Regional Names :

Eng. : The White Water Lily;
Kash : Birmposh, Kumud, Nilofer;
Mani. : Tharo.

Distribution : Trans-N.W.Himalaya (Lakes of Kashmir).
Part Used : Rhizomes.
Folk Uses : Rhizomes used as salad and food at the time of scarcity.
Active Constituents : Rhizomes comprise of ash, 10.8; carbohydrates (starch), 46; fibres, 10.0 and proteins, 6.4%.

References : Agarwal (2003), Ambasta (1986), Anonymous (1948-1976), Bennet *et al.* (1991), Raju (2000), Rana (2004), Rana *et al.* (2005), Singh *et al.* (2001b), Watt (1889-1896).

Nymphaea lotus Linn.

Syn : *N. edulis* Roxb.; *N. esculenta* Roxb.; *N. pubiscens* Willd.; *N. rubra* Roxb.

Family : Nymphaeaceae.

English, Hindi and Sanskrit Names:

Eng. : Indian Lotus;
Hindi : Kamal, Kanval;
Sans. : Hallaka, Kalharamu, Kamala, Kumda.

Distribution : Common throughout the wetlands of India.

Part Used : Roots.

Folk Uses : Roots contain a large quantity of starch and is usually baked or boiled with salt though sometimes eaten raw. These are also used as a famine food and eaten after roasting in ashes or dried before being grounded into flour.

Active Constituents : Chemical composition (g/100g) of root reveals ash, 6.32; carbohydrates (soluble), 66.78; fats, 7.8; fibres, 5.81; proteins, 9.3; and water, 10.94%.

References : Gammie (1902), Das *et al.* (1996), Saini (1996), Shortt (1887-1888), Watt (1889-1896).

Nymphaea microantha Guill & Perr.

Family : Nymphaeaceae.

Hindi Name :

Hindi : Nilkamal.

Distribution : N.E.Himalaya.

Part Used : Rhizomatous Roots.

Folk Uses : The underground roots consumed as

vegetable after boiling. The 'Shingju' is also prepared from it.

Reference : Singh *et al.* (2001b).

Nymphaea nauchali Burn. f.

Syn : *N. lotus* Hook.Thoms.non Linn.; *N. rubra ex* Salisb.

Family : Nymphaeaceae.

English, Hindi and Regional Names:.

Eng.	:	Indian Red Water Lily;
Hindi	:	Bhengat, Kanval, Kooka, Kooi ;
Ass.	:	Mokula, Nal;
Beng.	:	Nal, Rakto, Shaluk;
Guj.	:	Kanval, Nilophal;
Mal	:	Parimbal;
Mani.	:	Bhe, Thariktha- anganaba;
Kan.	:	Nadale-huvu;
Tel.	:	Allitamara.

Distribution : N.E.Himalaya and Wetlands of India

Reproductive Cycle : Throughout the year.

Part Used : Rhizomes.

Folk Uses : Rhizomes contain a large amount of starch and usually baked or boiled with water and eaten with salt. Also cooked in curries.

Active Constituents : Fresh rhizome comprises of carbohydrates (starch), 27.7; carbohydrates, 9.07; fats, 1.06; moisture, 53.95; proteins, 5.87 % and minerals.

References : Agarwal (2003), Ambasta (1986), Anonymous (1948-1976), Arora (1981), Arora & Pandey (1996), Bennet *et al.* (1991), Borthakur (1996), Singh *et al.* (2001b), Watt (1889-1896).

Nymphaea stellata Willd.

Family : Nymphaeaceae.

English, Hindi, Sanskrit and Regional Names:

Eng.	:	Indian Blue Water Lily;
Hindi	:	Nilkamal, Nil Padam;
Sans.	:	Indivera, Nilotpala, Nilpadma, Utpala;

Beng.	:	Nalla-Kalava;
Guj.	:	Nilkamal;
Mar.	:	Krishnakamal;
Mal	:	Sitambal;
Mani.	:	Thariktha;
Ori	:	Sudhiken;
Pub.	:	Banbher;
Tam.	:	Nilatpalam;
Tel.	:	Nallakalava.

Distribution : Common throughout the warmer regions of India.

Reproductive Cycle : August-November.

Parts Used : Rhizomes. Roots.

Folk Uses : Egg-sized rhizomes eaten as vegetable and consumed after boiling or roasting. Roots are also eaten.

Active Constituents : Dried rhizome contains carbohydrates, 67.49; fats, 0.45 ; moisture, 4.20; proteins, 14.15% and minerals.

References : Agarwal (2003), Ambasta (1986), Anonymous (1948-1976), Arora & Pandey (1996), Bennet *et al.* (1991), Borthakur (1996), Raju (2000), Samant *et al.* (2001), Singh *et al.* (2001b), Watt (1889-1896).

Nymphoides cristata (Roxb.) O. Kuntze.

Syn : *Limnanthemum cristatum* Grisb.

Family : Gentianaceae.

Hindi and Regional Names :

Hindi	:	Ghainchu, Kumuda, Tagarmul;
Mani.	:	Tharomacha.

Distribution : N.E.Himalaya.

Part Used : Rhizomes.

Folk Use : Rhizomes edible.

Reference : Singh *et al.* (2001b).

Nymphoides hydrophyllum (Lour.) O. Kuntze.

Family : Gentianaceae.

Regional Name :

Mani. : Ngachak -komal.

Distribution : Wetlands of N.E. Himalaya.

Part Used : Rootstocks.

Folk Use : The underground rootstocks consumed after boiling.

Reference : Singh *et al.* (2001b).

Nymphoides indicum (Linn.) Kuntze.

Syn : *Limenthemum indicum* (Linn.) Thaw.

Family : Gentianaceae.

English, Hindi and Regional Names:

Eng. : White Lotus;
Hindi : Bara-chuli;
Mani. : Nagachok -komal.

Distribution : Throughout India in tanks and ditches.

Reproductive Cycle : June- September.

Parts Used : Rhizomes. Roots. Stolons.

Folk Uses : Rhizome, roots and stolons used in soups, curries and as salad. The underground rootstocks also consumed after boiling.

References : Agarwal (2003), Arora (1981), Bennet *et al.* (1991), Borthakur (1996), Raju (2000), Samant *et al.* (2001), Singh *et al.* (2001b).

Onosma echinoides C.B.Clarke non Linn.

Syn : *O. hispidium* Wall. *ex* D.Don.

Family : Boranginaceae.

Hindi Names :

Hindi : Ratanjot.

Distribution : N.W. Himalaya (Kashmir and H.P.): 3,000 -3,800 m.

Part Used : Roots.

Folk Use : Roots yield edible dye.

References : Anonymous (1948-1976), Bennet *et al.* (1991).

Orthosiphon rubicundus Benth.

Family : Lamiaceae.

Hindi Name :

Hindi : Jikipota.

Distribution : N.W. Himalaya (Kashmir to Nepal). Deccan. W. Ghats upto 1800m.

Part Used : Tuberous Fusiform Roots.

Folk Use : Starchy tubers edible.

References : Anonymous (1948-1976), Arora & Pandey (1996), Rana *et al.* (2005).

Oxalis martiana Zuec.

Syn : *O. corymbosa* DC.

Family : Oxalidaceae.

Hindi and Regional Names:

Hindi : Khatta-mitha;
Del. : Khatmithi;
Tam. : Persia-Puliyarai.

Distribution : Almost throughout India.

Part Used : Tuberous Bulbs.

Folk Use : Small tuberous bulbs fragrant and eaten raw.

Active Constituents : Chemical composition of tuberous bulbs shows minerals, 2.0 (Ca, 0.41; K, 0.98; P, 0.18 mg/100g,); moisture, 74.4; nitrogenous compound, 0.7% and starch.

References : Agarwal (2003), Arora (1981, 1985a), Arora & Pandey (1996).

Pachyrhizus erosus (Linn.) Urban.

Syn : *Dolichos bulbosus* Linn.; *P. angulatus* Rich. *ex* DC.; *P. bulbous* Kurz.; *P. trilobus* DC.

Family : Papilionaceae.

English and Regional Names:

Eng. : Yam Bean, Yoka , Wayaka;
Beng. : Sankalu.

Distribution : Almost throughout India.
Part Used : Tubers.
Folk Uses : Yam tubers crisp, juicy, refreshing and eaten raw or cooked or sliced and made into chips. When cooked it is dirty white in colour and has an insipid flavour but palatable enough and used at the time of food scarcity.
Active Constituents : Peeled tubers comprise of ash, 0.50; carbohydrates (starch, 7.72; nonreducing sugars, 3.03); fats, 0.09; fibres, 0.64; minerals, 1.0 (Ca, 16.0; Cu, 0.43; Fe, 1.13mg/100g); moisture, 82.38; proteins, 1.47%; riboflavin, 0.02; thiamine, 0.05; niacin, 0.02 and vit. C (ascorbic acid), 14mg /100g.
References : Agarwal (2003), Ambasta (1986), Annonymous (1948-1976), Watt (1889-1896).

Pachyrhizus tuberosus Spreng.

Family : Papilionaceae.

Distribution : N. E.& N.W.Himalaya, Often cultivated.
Part Used : Tubers.
Folk Uses : Plant often cultivated for its nutritious tubers. Starch obtained from the tubers used in pudding, confectionery and bakery items.
References : Ambasta (1986), Anonymous (1948-1976).

Paeonia emodi Wall. *ex* Royle

Family : Ranunculaceae.

English and Regional Names:
Eng. : Himalayan Peony;
Kash. : Mid;
Pub. : Chandra, Mamekh.

Distribution : N.W. Himalaya (Kashmir to Kumaon): 2,600-3,000 m. Also grown in gardens.

Part Used : Tubers

Folk Uses : Tubers eaten after repeated boiling.

References : Anonymous (1948-1976), Bennet *et al.* (1991).

Remarks : Tubers considered poisonous and should be eaten after proper boiling.

Paris polyphylla Smith

Family : Liliaceae.

Distribution : N.W. Himalaya (Kumaon region): 2,000 - 2,800 m.

Part Used : Roots.

Folk Use : Roots edible.

References : Samant *et al.* (2001).

Passiflora quadrangularis Linn.

Family : Passifloriaceae.

English Name :

Eng. : True or Giant Grandiila.

Distribution : Grown in Indian gardens as an ornamental plant.

Part Used : Tuberous Roots.

Folk Use : Tuberous roots eaten like Yams.

References : Ambasta (1986), Anonymous (1948-1976).

Pastinaca sativa Linn.

Syn : *Peucedanum sativum* Benth. & Hook.f.

Family : Apiaceae.

English Name :

Eng. : Parsnip.

Distribution : Colder regions of India (cultivated).

Part Used : Fleshy Tap Roots.

Folk Uses : Fleshy roots known as 'Parsnips' are consumed as vegetable and salad, and also employed in soups and wines.

Active Constituents : Roots exihibit ash, 0.6-1.5; carbohydrates, 23.2; fats, 0.3; fibres, 1.7; minerals, 1.1 (Ca,

50; P, 40 ; Fe, 0.4mg/100g); moisture, 72.4; proteins 1.3%; vit. A, 301 I.U.; niacin, 0.4; riboflavin, 0.06-0.12; thiamine, 0.31mg/100g and vit. C, 16 /100g.

References : Agarwal (2003), Ambasta (1986), Anonymous (1948-1976).

Pentatropis spiralis Decne.

Syn : *P. cynachoides* R.Br.

Family : Ascelpiadaceae.

Hindi Sanskrit and Regional Names:

Hindi : Amberbel, Kauathodi;
Guj. : Shigaroti;
Mar : Shvetkavali;
Pub. : Ambravel, Vanvari.

Distribution : W. Himalaya. Gangetic plains, Punjab, Delhi. Gujarat (Saurashtra).

Reproductive Cycle : January- October.

Part Used : Tubers.

Folk Use : Sweet tubers eaten after peeling.

References : Agarwal (2003), Ambasta (1986), Anonymous (1948-1976), Arora & Pandey (1996).

Persea macroantha Mill.

Family : Lauraceae.

Distribution : Himalaya.

Part Used : Tubers.

Folk Use : Tubers edible.

Reference : Krishna *et al.* (1999).

Peucedanum dhana Buch-Ham *ex* C.B.Clark

Family : Apiaceae.

Regional Names :

Bih. : Bhojraj, Mann Tiro;
Bomb. : Koland.

Distribution : W. Himalaya upto 2,300 m. Maharashtra. Bihar. Orissa.

Reproductive Cycle : March -June.
Part Used : Roots.
Folk Use : Swollen roots taste like carrot and eaten raw and considered as tonic.
References : Arora & Pandey (1996), Bennet *et al.* (1991).

Phaseolus adnanthus G.F.W. Mey.

Family : Papillionaceae.

Hindi and Regional Names:

Hindi : Banbarbatti;
Beng. : Banbarti;
Bomb. : Kullonda;
Mal. : Kattupayaru;
Tam. : Kattupayaru;
Tel. : Karalasana.

Distribution : Himalayan terrain to the plains of India. Cultivated as an ornamental plant.
Part Used : Tuberous Roots.
Folk Use : Tuberous roots eaten as food at the time of scarcity.
References : Agarwal (2003), Ambasta (1986), Anonymous (1948-1976), Watt (1889-1896).

Phaseolus rostratus Wall.

Family : Papilionaceae.

Regional Names :

Tam. : Karlsona;
Tel. : Karlasana, Karualachandra.

Distribution : Maharashtra.
Part Used : Tuberous Roots.
Folk Use : Tuberous roots eaten cooked.
Reference : Shortt (1887-1888).

Phragmites communis Trin.

Family : Poaceae.

English, Hindi and Regional Names:

Eng. : The Common Reed;
Hindi : Dila, Damu;
Pub. : Dila, Damba.

Distribution : N.W.Himalaya from Kashmir to Kumaon.
Parts Used : Rhizomes. Roots
Folk Use : Rhizomes and roots edible
Active Constituents : Rhizomes rich in carbohydrates.
References : Anonymous (1948-1976), Arora & Pandey (1996).

Phyllostachys bambusoides Sieb. & Zuc.

Family : Poaceae.

Hindi Name :

Hindi : Tabosheer- ban.

Distribution : E. Himalaya (Assam).
Part Used : Roots.
Folk Use : Roots edible.
References : Agarwal (2003), Anonymous (1948-1976).

Pleurospermum angilicoides (DC.) Cl.

Family : Apiaceae.

Distribution : C. Himalaya: 2,500- 3,200 m.
Part Used : Roots.
Folk Use : Roots edible.
Reference : Samant *et al.* (2001).

Plesmonium margiritiferum Schott.

Family : Araceae.

English, Hindi and Regional Names:

Hindi : Kharkar;
Beng. : Gajeer-mul;
M.P. : Bansur, Jaluliya;
Mund. : Birkanda.

Distribution : Throughout C. and S. India. Bihar. C. W. Bengal. M.P.
Part Used : Tubers.
Folk Use : Tubers eaten after proper boiling in tamarind.
References : Agarwal (2003), Anonymous (1948-1976).

Remarks : Tubers considered toxic and should be consumed after boiling in tamarind.

Polygonatum cirrhifolium (Wall.) Royle

Family : Liliaceae.

Distribution : N.W. Himalaya (Kumaon to Garhwal): 2,500-3,300 m.

Part Used : Tubers.

Folk Use : Tubers edible.

References : Rana *et al.* (2005), Samant *et al.* (2001).

Polygonatum multiflorum All.

Syn : *Canvallaria multilora* Linn.; *C. govaniana* Wall.

Family : Liliaceae.

English Name :

Eng. : Soloman Seal.

Distribution : Trans-N.W. and E. Himalaya (J. & K. to Kumaon). Manipur: 1,800-2,700 m.

Part Used : Rhizomes.

Folk Uses : Rhizomes edible and its pieces cooked in curries, also used in the preparation of thick soup by tribals of Manipur.

References : Anonymous (1948-1976), Arora (1981), Arora & Pandey (1996), Sinha (1996), Watt (1889-1896).

Polygonatum oppositifolium (Wall.) Royle

Family : Liliaceae.

Regional Name:

Sikk. : Garimbosa.

Distribution : E. and N.E. Himalaya (Nepal to Assam). Khasia Hills: 1,200- 2,000 m.

Reproductive Cycle : July- September.

Part Used : Roots.

Folk Use : Roots edible especially in N.E. states.

References : Bennet *et al.* (1991), Chandra (1997), Sinha (1996).

Polygonatum verticillatum All.

Syn : *Campydorum verticillatum* Salisb.; *Canvellaria verticillata* Linn.; *Canvellaria lepitophylla* D. Don.; *Polygonatum jacquemonatium* Kunth.; *P. lepidophyllum* Royle.

Family : Liliaceae.

Hindi and Regional Names:

Hindi : Meetha-dudia, Shakakal misri;
Pub. : Shakakul.

Distribution : Temperate Himalaya (Kashmir eastwards to Manipur): 2,700-4,000m.

Part Used : Rhizomes.

Folk Uses : Rhizomes valued as strength giving food in Kumaon and Garhwal regions. Starchy rhizomes also cooked in curries and valued in the preparation of thick soup by Manipuri tribal.

References : Agarwal (2003), Ambasta (1986), Arora & Pandey (1996), Bennet *et al.* (1991), Rana *et al.* (2005), Sinha (1996), Samant et al. (2001), Watt (1889-1896).

Polygonum affine Don.

Family : Polygonaceae.

Distribution : C. and N.W. Himalaya (Kumaon and Garhwal to Sikkim): 3,000-4,000 m.

Part Used : Roots.

Folk Use : Roots edible and eaten by locals.

Reference : Samant *et al.* (2001).

Polygonum alatum Ham. var. nepalensis

Syn : *P.nepalensis* Meissn.; *P.guttuliferum* Miq.

Family : Polygonaceae.

Regional Names :

Megh. : Jaut;
Pub. : Sat-halon.

Distribution : Throughout the Himalaya (Kashmir to Sikkim): 2,000-3300 m.

Part Used : Roots.
Folk Use : Roots edible.
References : Joseph & Kharkongor (1996), Samant et al. (2001), Watt (1889-1896).

Polygonum amplexicaule D.Don.

Family : Polygonaceae.

Regional Names :

U.P. : Durpa-tander, Kakur-makri.

Distribution : Entire Himalaya (Kashmir to Sikkim) upto 5,000 m.
Reproductive Cycle : June-October.
Parts Used : Rhizomes. Roots
Folk Uses : Rhizomes used as a substitute for tea leaves and also recommended as body tonic. Roots sold as 'Anjubar' in the markets of H.P.
References : Chandra (1997), Pangtey *et al.* (1982), Samant *et al.* (2001).

Polygonum aviculare Linn.

Family : Polygonaceae.

English, Hindi, Sanskrit and Regional Names:

Eng. : The Knot Grass;
Hindi : Ban-nalia, Ban-natia, Hunraj, Machoti;
Sans. : Miramoti. Nisamoli;
Beng. : Machuti;
Kash. : Drop;
Pub. : Banduke, Kesru.

Distribution : N.W.Himalaya (Kashmir to Kumaon): 2,000-3,500 m.
Reproductive Cycle : June- October.
Part Used : Roots.
Folk Use : Roots used as vegetable and also added to curries.
References : Agarwal (2003), Chandra (1997), Rana (2004), Watt (1889-1896).

Polygonum bistorta Linn.

Syn : *Polygonum paleaceum* Wall. *ex* Hook.f.

Family : Polygonaceae.

English and Hindi Names:

Eng. : Bistort-root;
Hindi : Anjubar.

Distribution : E. and N.W. Himalaya (Kashmir, Kumaon, Garhwal to Assam): 2,700-4,500 m.

Parts Used : Roots. Rootstocks.

Folk Uses : Rootstocks eaten boiled and as vegetable as well as used in soups, stews, curries, etc, and constitute the 'Anjubar' of Arabs. Thick roots rich in starch are dried and grounded into flour for making bread.

References : Agarwal (1986, 2003), Ambasta (1986), Arora (1981), Arora & Pandey (1996), Anonymous (1948-1976), Chandra (1997), Samant *et al.* (2001).

Polygonum chinense Linn.

Family : Polygonaceae.

Regional Names:

Ass. : Kelnap, Kuki;
Mar. : Paral;
Megh. : Jallynoh;
Kan. : Bilichini-gangalu.

Distribution : N.E.Himalaya and almost all over India upto 3,100 m.

Reproductive Cycle : January- May.

Part Used : Rootstocks.

Folk Use : Rootstocks eaten by N.E. tribals especially in Khasia, Jantia and in Lakhimpur.

References : Agarwal (2003), Joseph & Kharkongar (1996), Chandra (1997), Samant *et al.* (2001).

Polygonum glabrum Willd.

Syn : *P. persicaria* Wall.; *P.poiretii* Meissn.

Family : Polygonaceae.

Regional Names :

Ass.	:	Behlangani, Larbourna, Pathurva ;
Beng.	:	Bhigni;
Kan.	:	Nirukanigalu-latkta;
Mal.	:	Chuvanna-mudela-mukku;
Tam.	:	Actalaree.

Distribution	:	Generally distributed throughout India.
Reproductive Cycle	:	September-February (Next year).
Part Used	:	Roots.
Folk Use	:	Roots edible and used as vegetable.
References	:	Bennet *et al.* (1991), Rana (2004), Vartak (1981), Watt (1889-1896).

Polygonum multiflorum All.

Family : Polygonaceae.

English and Hindi Names:

Eng.	:	Soloman's Seal ;
Hindi	:	Seal-mul.

Distribution	:	E. and N.W. Himalaya (Kashmir, Kumaon, Garhwal, Manipur) upto 3,000 m. Also cultivated in Indian gardens as an ornamental plant.
Parts Used	:	Roots. Rootstocks.
Folk Uses	:	Rootstocks used as vegetable. Roots soaked overnight and boiled in changes of water to remove the bitter principle before eating.
References	:	Agarwal (2003), Ambasta (1986), Bennet *et al.* (1991).

Polygonum plebium R.Br.

Family : Polygonaceae.

Regional Name :

Mani.	:	Tarakmana.

Distribution	:	Chiefly E. Himalaya (Manipur). Greater parts of India.

Part Used : Roots.
Folk Use : Roots used as vegetable.
References : Agarwal (2003), Rana (2004), Sinha (1996).

Polygonum viviparum Linn.

Family : Polygonaceae.

English, Hindi and Regional Names:

Eng. : Viviparus Bistort;
Hindi : Maslun, Pahari-chemeti,
Pub. : Bajir, Balisuri, Mamecha, Maslun.

Distribution : Entire Himalaya.
Part Used : Rootstocks.
Folk Use : Rootstocks edible, and also recommended as tonic.
References : Agarwal (2003), Anonymous (1948-1976), Arora & Pandey (1996).

Portulaca grandifolia Hook.f.

Family : Portulacaceae.

English Name :

Eng. : Rose-moss.

Distribution : Warmer regions of India.
Part Used : Roots.
Folk Use : Roots used as vegetable.
Reference : Agarwal (2003).

Portulaca pilosa Linn.

Syn : *P. tuberosa* Roxb.
Family : Portulacaceae.

Regional Names :

Beng. : Loniya;
Mar. : Jangli Gajar;
Tel. : Bodda -kura.

Distribution : C. and W. India and in Deccan Peninsula; mostly in warmer regions of India.
Part Used : Fusiform Tap Roots.

Folk Uses : Roots eaten raw by shepherd's boys. Roots used as vegetable.

Active Constituents : Roots rich in starch and iron.

References : Agarwal (2003), Chandra (1997), Saxena (1979).

Portulaca quadrifolia Linn.

Family : Portulacaceae.

Hindi, Sanskrit and Regional Names:

Hindi : Chounlayi, Khate-chaval;
Sans. : Uppadykii;
Beng. : Chotta-luniya, Nunia;
Guj. : Luni;
Mar. : Khate-chanval;
Kan. : Gunni-sopu;
Tam. : Chinna-parpukkirai;
Tel. : Palaya-kura.

Distribution : Warmer regions of India.

Part Used : Roots.

Folk Use : Roots used as vegetable.

References : Agarwal (2003), Anonymous (1948-1976), Gammie (1902), Gupta & Kanodia (1968), Saxena (1979), Shortt (1887-1888).

Potamogeton crispus Linn.

Family : Potamogetonaceae.

Distribution : Grown all over India

Part Used : Rootstocks.

Folk Uses : Rootstock is edible and eaten raw or cooked.

Reference : Agarwal (2003).

Potamogeton filiformis Pers.

Family : Potamogetonaceae.

Distribution : Found throughout India in wetlands.

Part Used : Rootstocks.

Folk Uses : Rootstocks contain starch which is extracted and edible.

Active Constituents : Analysis of rootstock reveals the occurrence of carbohydrates, 7.0; cellulose, 19.1-23.7; fats, 0.6-1.9 ; fibres, 0.6-2.5; nitrogen free extract, 45.5-52.2; minerals, 16.2-18.3 and proteins (crude), 7.6-15.2%.

References : Ambasta (1986), Anonymous (1948-1976), Arora & Pandey (1996).

Potamogeton graminis Linn.

Family : Potamogetonaceae.

Distribution : Throughout India.

Part Used : Rootstocks.

Folk Use : Rootstocks contain edible starch.

References : Ambasta (1986), Arora & Pandey (1996).

Potamogeton pectinatus Linn.

Family : Potamogetonaceae.

Distribution : Wetlands all over India.

Part Used : Rootstocks.

Folk Uses : Rootstocks contain edible starch.

Active Constituents : Rootstock on dry matter basis comprises of carbohydrates, 6.0; cellulose, 17.1-22.7; fats, 0.6-1.9; fibres, 0.6-2.5; nitrogen free extract, 44.5-51.2; minerals, 9.2-11.2 and proteins (crude), 6.5-15.3%.

References : Ambasta (1986), Anonymous (1948-1976), Arora & Pandey (1996).

Potamogeton perfoliatus Linn.

Family : Potamogetonaceae.

Distribution : Temperate and tropical Himalaya.

Part Used : Rhizomes. Roots.

Folk Use : Both rhizomes and roots are edible and contain starch.

Active Constituents : Chemical analysis of roots and rhizome on dry matter basis shows carbohydrates, 6.0; cellulose, 13.8-15.9; fats, 0.6-1.3 ;

fibres, 0.6-2.5; nitrogen free extract, 45.5-59.2; minerals, 12.3-31.5 and proteins (crude), 6.6-13.3%.

Reference : Anonymous (1948-1976).

Potentilla anserina Linn.

Family : Rosaceae.

English and Hindi Names:

Eng. : Silver weeds;
Hindi : Penma.

Distribution : W. Himalaya: 2,100- 4,800 m. Grown on rocky soil in Indian gardens.

Part Used : Rootstocks.

Folk Uses : Rootstocks eaten at the time of scarcity.

References : Agarwal (2003), Ambasta (1986), Anonymous (1948-1976).

Potentilla fulgens Hook.

Family : Rosaceae.

Regional Names :

Beng. : Buitara;
Megh. : Lyngiangbru (Khasi Hills).

Distribution : W. temperate Himalaya extending eastwards to Khasia in Meghalaya.

Part Used : Roots.

Folk Use : Roots eaten as arecanut alongwith betel leaves.

Reference : Arora & Pandey (1996).

Pouzolzia tuberosa Wight.

Syn : *P. indica* Gaudich.; *P. zelaninca* (Linn.) Benn.; *Urtica tuberosa* Roxb.

Family : Urticaceae.

Hindi and Regional Names:

Hindi : Dumpa;
Mal. : Kullu-raki;
Tel. : Eddu.

Distribution : Found throughout India, chiefly in S. India.

Reproductive Cycle : October-January (Next year).

Part Used : Root Tubers.

Folk Uses : Tuberous roots eaten raw, boiled or roasted alongwith its leaves.

References : Agarwal (2003), Ambasta (1986), Anonymous (1948-1976), Arora & Pandey (1996), Bennet *et al.* (1991), Rana (2004), Watt (1889-1896).

Pteridium aquilinium (Linn.) Kuhn.

Syn : *Pteris aquiline* Linn.
Family : Polypodiaceae.

English, Hindi, Sanskrit, and Regional Names:

Eng. : Brake or Braken Fern;
Hindi : Brakaparni.

Distribution : Throughout India upto 3,600m. Also grown as an ornamental plant.

Part Used : Rhizomes.

Folk Uses : Rhizomes boiled or roasted and eaten at the time of scarcity, or grounded into flour for making bread though its bitterness is removed by repeated washing. For brewing beverages it is mixed with malt. Also the powdered starch is used in bakery.

Active Constituents : Analysis of dried rhizomes on dry matter basis reveals carbohydrates, 51.0; fats, 1.2; fibres, 20.0; proteins, 9.5; moisture, 10.0 and ash 8.3%

References : Agarwal (2003), Ambasta (1986), Bennet *et al.* (1991), Hosagouder & Henrey (1996), Singh & Viswanathan (1996), Vartak (1981).

Pueraria lobota (Willd) Ohwi.

Syn : *P. thunbergiana* (Sieb. & Zucc.) Benth.;
P. hirusta Schneid. non Kurz. Kudzu.
Family : Papilionaceae.

English and Hindi Names:

Eng. : Kudzu;
Hindi : Termei.

Distribution : Tropical regions of India.
Part Used : Tuberous Roots.
Folk Uses : Tuberous roots eaten as vegetable and its flour used in soups and prescribed for general debility.
Active Constituents : Roots contain ash, 1.4; carbohydrates, 27.8; fats, 0.1 ; fibres, 0.7; minerals, 1.0(Ca, 15; P, 18; Fe, 0.6mg/100g); moisture, 68.6; proteins, 2.1%; didzin glucoside and purarin isoflavanoid.
References : Agarwal (2003), Anonymous (1948-1976).

Pueraria phaseoloides Benth.

Syn : *P. javanica* Benth.
Family : Papilionaceae.

Hindi and Regional Names:

Hindi : Kudzu ;
Ass. : Jermai-kyn-saw,Jermei-so-gonosh.

Distribution : Subtropical Himalaya.
Part Used : Roots.
Folk Use : Root tubers edible.
References : Agarwal (2003), Arora & Pandey (1996).

Pueraria thomsonii Benth.

Family : Papilionaceae.

Regional Name :

Ass. : Kai-kuangra, Suiting.

Distribution : N.E. and Subtropical Himalaya chiefly in Assam, Khasia and Lushai Hills.

Part Used : Root Tubers.

Folk Use : Root tubers edible.

References : Agarwal (2003), Arora & Pandey (1996).

Pueraria tuberosa Roxb. *ex* Benth. (Pl. 8A)

Syn : *Hedysarum tuberosum* Roxb.

Family : Papilionaceae.

English, Hindi, Sanskrit and Regional Names:

Eng. : Indian Kudzu;
Hindi : Bidari-kand, Bilai-kand, Bharda, Kohnda, Sural, Tirra;
Sans. : Bankumara;
Beng. : Batraji, Shimia;
Guj. : Khakervel, Phagvelo, Vadarikand;
Kan. : Gummadigida;
Mar. : Ghorbel;
Pub. : Bidari-kand;
Tel. : Darigummadi.

Distribution : E. and C. Himalaya and generally distributed throughout India: 300-1,200 m.

Reproductive Cycle : February-June.

Part Used : Root Tubers.

Folk Uses : Yam-like root tubers weighing upto 40 Kg each, tasted like liquorice roots and eaten raw or boiled or after roasting. Also used for extraction of starch.

Active Constituents : Tubers comprise of (dry matter, 85.5; moisture, 14.1) carbohydrates, 64.6; crude fibres, 28.8; crude proteins, 10.9; fats, 0.3; glucose, fructose and sucrose.

References : Agarwal (2003), Ambasta (1986), Anonymous (1948-1976), Arora & Pandey (1996), Bennet *et al.* (1991), Rana (2004), Rao & Henery (1996), Samant *et al.* (2001), Singh and Anand (1996), Watt (1889-1896).

Raphanus sativus Linn.

Family : Brassicaceae.

English, Hindi, Sanskrit and Regional Names:

Eng. : The Raddish;
Hindi : Muli, Muro;
Sans. : Mulika;
Beng. : Mulla;
Pub. : Mungra , Muli, Tara-mira;
Tam. : Mulangi;
Tel. : Mulangi.

Distribution : Cultivated all over India.

Reproductive Cycle : July–December.

Part Used : Fleshy Tap Roots.

Folk Uses : Roots eaten raw or as vegetable. Also it is used as salad and for pickling.

Active Constituents : White variety shows edible matter, 99.0; carbohydrates, 3.4; fats, 0.1 ; fibres, 0.8 ; minerals, 0.6 %, (Ca,50.0 ; P, 22.0; Fe, 0.4 mg/100g,); moisture, 94.4; proteins 0.7%; vit.A, 5.0 I.U.; riboflavin, 0.02mg/100g; thiamine, 0.06mg/100g; nicotinic acid, 0.05mg/100g; vit. C, 15.0mg /100g and that of violet variety reveals edible matter, 94.0; out of which carbohydrates form, 6.8; fats, 0.3 ; fibres, 0.6 ; minerals, 0.9 %, (Ca, 50.0 ; P, 20.0; Fe, 0.5 mg/100g,); moisture, 90.8; proteins, 0.6 %; vit. A, 5.0 I.U.; riboflavin, 0.02mg/100g; thiamine, 0.06mg/100g; nicotinic acid, 0.04mg/100g and vit. C, 17.0mg /100g.

References : Agarwal (2003), Ambasta (1986), Anonymous (1948-1976), Watt (1889-1896).

Raphanus sativus Linn. var. *caudatus*

Family : Brassicaceae.

English and Regional Names:

Eng. : The Rat Tailed Raddish;
Pub. : Mungra.

Distribution : Cultivated in Punjab and Haryana.
Reproductive Cycle : July -December.
Part Used : Roots.
Folk Use : Roots edible but chiefly grown for its seeds and pods.
References : Ambasta (1986), Watt (1889-1896).

Raphidophora lacinata (Burm.f.) Merr.

Family : Araceae.

Hindi Name :

Hindi : Ganesh-kanda.

Distribution : Throughout S. India.
Part Used : Rhizomes.
Folk Uses : Rhizomes eaten after repeated boiling and washing.
Reference : Agarwal (2003).
Remarks : Rhizomes considered as poisonous and should be consumed only after proper boiling.

Remustia vivipera (Lodd.) Schutt.

Family : Araceae.

Regional Name :

Mar. : Garget, Rukhalu.

Distribution : Tropical and Subtropical Himalaya. Maharashtra. Goa .
Part Used : Corms.
Folk Uses : Tuberous corms edible but after repeated boiling to rid off irritating crystals of calcium oxalate.
References : Ambasta (1986), Anonymous (1948-1976), Arora & Pandey (1996), Bennet *et al.* (1991), Vartak (1981).
Remarks : Corm to be consumed only after careful boiling.

Rheum raphonticum Linn.

Family : Polygonaceae.

English and Hindi Names:

Eng. : Rhapontic Rhubarb;
Hindi : Bag ki Rubarb.

Distribution : Cultivated all over India.

Parts Used : Rhizomes. Roots.

Folk Uses : Rhizomatous roots alongwith petiole constitute the 'Rhubarb' and considered to possess high proteins content and are edible.

Reference : Agarwal (2003).

Rumex patientia Linn.

Syn : *R. orientalis* Bernh.
Family : Polygonaceae.

English Name :

Eng. : Patience Dock.

Distribution : W.Himalaya.

Part Used : Roots.

Folk Use : Roots sweet and eaten raw.

References : Ambasta (1986), Arora & Pandey (1996).

Sagittaria sagittaefolia (Linn.) DC.

Family : Alismataceae.

Regional Name :

Ass. : Kauka.

Distribution : Throughout India and N. E. Himalaya.

Part Used : Corms.

Folk Uses : Fleshy corms eaten boiled by the natives, and also used in the preparation of 'Iromba'.

Active Constituents : Analysis of corm reveals ash, 1.6; carbohydrates, 23.7; fats, 0.2; fibres, 0.8; minerals, 1.0 (Ca, 16; P, 216; Fe,1.4 mg/

100g,); moisture, 69.5; proteins, 5.0%; and vit. C, 2mg /100g.

References : Arora (1981), Anonymous (1948-1976), Bennet *et al.* (1991), Singh *et al.* (2001b), Watt (1889-1896).

Remarks : It is slightly acrid and supposed to be poisonous by the natives and to be used only after proper boiling.

Sagittaria sinensis Linn.

Family : Alismataceae.

Distribution : N.E. Himalaya.

Part Used : Rhizomes.

Folk Uses : Tuberous rhizomes used in the preparation of 'Iromba', and also consumed in salad.

Reference : Singh *et al.* (2001b).

Sagittaria trifolia Linn.

Syn : *S. sagittifolia* Auct. non Linn.

Family : Alismataceae.

English, Hindi and Regional Names:

Eng. : Arrowhead;

Hindi : Chotakut;

Beng. : Chottokut, Muya-muja.

Distribution : Throughout India upto 1,800m.

Reproductive Cycle : October-February.

Part Used : Tubers.

Folk Uses : Tubers give an unpleasant taste due to milky juice but made into a sweetish palatable vegetable by cooking it after boiling or roasting. Also boiled tubers can be stored for future use after proper drying.

Active Constituents : Chemical analysis of tuber shows carbohydrates, 23.7; fats, 0.2 ; minerals, 1.0 (Ca,16.0; P, 26.1mg/100g,); moisture, 69.5 and proteins, 5.0 %.

References : Arora & Pandey (1996), Bennet *et al.* (1991).

Sauromatum venosum (Ait.) Kunth.

Syn : *S. guttatum* Schott.; *S. pedatum* Schott.

Family : Araceae.

Hindi, Sanskrit and Regional Names:

Hindi : Murangai;
Sans. : Dhey, Nurki;
Bomb. : Divakanda, Loth;
Mar. : Mundari, Nukari, Pindi;
M.P. : Bhasma-kand;
U.P. : Dhey, Minkerivally.

Distribution : N. E. and W. Himalaya (Kashmir to Kumaon Hills). Gujarat. Karnataka and Indo-gangetic plains.

Part Used : Tubers.

Folk Uses : Tubers acrid but consumed like that of *Dioscorea* spp. These are also used as condiments in certain parts of India.

References : Agarwal (2003), Ambasta (1986), Anonymous (1948-1976), Arora & Pandey (1996), Chandra (1997), Nilegaonkar *et al.* (1985).

Remarks : To remove the acrid principle, tubers should be consumed after thorough boiling.

Scirpus dubius Roxb.

Family : Cyperaceae.

Distribution : Wetlands of Konkan, Goa, and Ahmedabad.

Part Used : Roots.

Folk Uses : Tuberous roots considered good as 'Yams' and eaten by the labour class and famine stricken people.

Reference : Watt (1889-1896).

Scirpus grossus Linn.f.

Syn : *S. kysoor* Roxb.; *S. grossus* Linn.f. var. *kysoor* Cl. non Roxb.

Family : Cyperaceae.

Hindi, Sanskrit and Regional Names:

Hindi : Kasaru, Kesur;
Sans. : Kasurka;
Beng. : Kasaru, Kesur;
Mar. : Jome kesari, Kasara, Mundari;
Pub. : Dila, Kesara;
Tel. : Gunda, Tunga.

Distribution : N.E. and N.W. Himalaya and throughout India upto 700m.

Part Used : Tubers.

Folk Uses : Tubers sweet and starchy and considered highly nutrititious and eaten raw or cooked. For making bread, tubers are dried after removing the dark cuticle and grounded into flour or mixed with the barley flour.

Active Constituents : Chemical composition (g/100g) of tubers reveals carbohydrates (soluble), 76.18g.; fats, 1.0g; fibres, 6.80g; ash, 3.40g; proteins, 4.12g; water, 8.50g; calories, 338 and calcium.

References : Agarwal (2003), Ambasta (1986), Arora & Pandey (1996), Bhargava (1959), Das *et al.* (1996), Vartak (1981).

Scirpus lacustris Linn.

Family : Cyperaceae.

English and Regional Names :

Eng. : Club Rush, The Great Bulrush;
H.P. : Kesaru.

Distribution : Wetlands in Ladakh Kashmir to Kumaon upto 4,250 m.

Part Used : Rhizomes.

Folk Uses : Rhizomes eaten and also made into bread.

Active Constituents : Rhizomes on dry matter basis reported to contain ash, 7.84; carbohydrates, 10.4; fats, 2.68; fibres, 40.0; nitrogen free extract, 43.4; proteins (crude) 6.10 % and vit. B complex.

References : Anonymous (1948-1976), Arora & Pandey (1996), Rana (2004), Rana et al. (2005).

Scirpus tuberosus Desf.

Syn : *S. maritimus* C.B. Clarke. non Linn.

Family : Cyperaceae.

English, Hindi and Regional Names:

Eng. : Seaclub-Rush;
Hindi : Murak;
Beng. : Mahat;
Pub. : Dila, Murak;
Tel. : Gurrapa-sakatunga.

Distribution : Found in marshes throughout India.

Part Used : Rhizomes.

Folk Uses : Rhizomes made into flour for use in times of scarcity.

Active Constituents : On dry matter basis rhizomes yield ash, 7.84; crude cellulose, 31.0; crude fats, 2.2 ; fibres, 40.0; nitrogen free extract, 46.0; proteins (crude) 10.3% and vit. B complex.

References : Agarwal (2003), Ambasta (1986), Arora & Pandey (1996).

Scleria biflora Roxb.

Family : Cyperaceae.

Distribution : E. Himalaya and Maharashtra.

Part Used : Roots.

Folk Uses : Roots smell like camphor and eaten with rice, and also used as spice and condiment.

Reference : Agarwal (2003).

Sechium edule Sw.

Family : Cucurbitaceae.

English, Hindi and Regional Names:

Eng. : Chayote, Chow-Chow;
Hindi : Squash Kaddu;
Kan. : Seema-badane;
Tam. : Seema-katti-rikkai.

Distribution : Tropical Himalaya and Subtropical regions of India.

Part Used : Roots.

Folk Uses : Roots used as substitute of arrowroot starch, and eaten after boiling or roasting.

References : Agarwal (2003), Ambasta (1986), Anonymous (1948-1976), Arora & Pandey (1996), Rana (2004), Rana *et al.* (2005), Watt (1889-1896).

Sedum rosea Scop. *ex* Sprange

Syn : *S. rhodiola* DC.

Family : Crassulaceae.

English and Regional Names:

Eng. : Rose Root;
Lahul : Shrolo.

Distribution : Trans-N.W. Himalaya from Kashmir to Kumaon Hills upto 5,000 m.

Parts Used : Rootstocks. Roots.

Folk Uses : Fleshy rootstocks with the odour of roses are cooked and eaten or pickled. Roots used as tonic.

References : Agarwal (2003), Ambasta (1986), Anonymous (1948-1976), Rana *et al.* (2005).

Sedum tibeticum Hk.f.& Th.

Family : Crassulaceae.

Distribution : Trans-N.W. Himalaya: 3,600-4,800m.

Part Used : Roots.

Folk Use : Roots edible.
References : Anonymous (1948-1976), Samant *et al.* (2001).

Silybum laciniatum Linn.

Family : Asteraceae.

English and Hindi Names:

Eng. : Milk thistle;
Hindi : Dudhi Genda.

Distribution : Trans-N.W. Himalaya (Kashmir) upto 2,500 m. Semicultivated in gardens as an ornamental plant.
Part Used : Roots.
Folk Use : Roots cooked and eaten.
Reference : Agarwal (2003).

Silybum marianum Gaertn.

Family : Asteraceae.

Distribution : Trans-N.W. Himalaya (Kashmir); semicultivated as an ornamental plant.
Part Used : Roots.
Folk Use : Roots eaten after boiling
Reference : Arora & Pandey (1996).

Sinobambusa elangans Kurz.

Family : Poaceae.

Hindi Name :

Hindi : Nakari.

Distribution : Assam and Naga Hills (E. Himalaya).
Part Used : Tender Rhizomes.
Folk Use : Tender rhizomatous culms are pickled and eaten.
Reference : Arora & Pandey (1996).

Sium latijugum C.B.Clarke

Family : Apiaceae.

Distribution : N.W. Himalaya.

Part Used : Roots.
Folk Use : Roots edible.
Reference : Arora & Pandey (1996).

Smilax megacarpa DC.

Syn : *S. extensa* Wall. *ex* Hk.f.
Family : Liliaceae.

Distribution : N.E. Himalaya.
Part Used : Rhizomes.
Folk Use : Rhizomes edible.
Reference : Arora & Pandey (1996).

Smilax ovalifolia Roxb.

Syn : *S. macrophylla* Roxb.
Family : Liliaceae.

Hindi and Regional Names:

Hindi : Chob-chini, Jangli -ushbah;
Beng. : Kumarka;
Kan. : Nirrubeta;
Mal. : Kaltamara;
Mar. : Ghyvel-ghutti;
Ori : Mitri;
Tam. : Malaittamarai;
Tel. : Konda-tamara.

Distribution : Throughout Tropical regions of India.
Part Used : Roots.
Folk Use : Roots powdered into flour for making breads.
References : Agarwal (2003), Anonymous (1948-1976).

Smilax zelanica Linn.

Family : Liliaceae.

Distribution : N.W. Himalaya (Kumaon region) upto 1,000 m.
Part Used : Tender Roots.
Folk Use : Roots edible.
References : Anonymous (1948-1976), Samant *et al.* (2001).

Solanum trilobatum Linn.

Syn : *S. acetosaefolium* Lamk.; *S. canaranum* Miq.
Family : Solanaceae.

Hindi Name :

Hindi : Alarka.

Distribution : Common in Deccan Peninsula.

Part Used : Roots.

Folk Use : Bitter roots edible.

References : Agarwal (2003), Watt (1889-1896).

Solanum tuberosum Linn. (Pl. 8C)

Family : Solanaceae.

English, Hindi and Regional Names:

Eng. : Potato;
Hindi : Alu;
Beng. : Alu, Batata;
Kan. : Batata;
Tel. : Alu, Gudalu, Utalaygudda.

Distribution : Cultivated all over India.

Part Used : Stem Tubers.

Folk Uses : Underground tubers widely consumed as food all over the world by almost all classes of people in the form of curries, chips, soups and as vegetable. More than hundred types of its products are manufactured and sold in markets all over the world. Also it is used in the manufacture of starch and beverages.

Active Constituents : Chemical analysis of Indian potato reveals (% of material) ash, 0.9-1.1; carbohydrates, 18.9-21.4; fats, 0.2; minerals (Ca, 6.5; P, 37.4; Fe, 1.4, Mg, 20; Na, 11.0; K, 247; Cu, 0.20; S, 37.7; Cl, 16 mg/100g); small quantity of I, Mn and Zn; moisture, 75.4-77.0; nitrogen, 1.2 and proteins (crude), 1.8-2.3%. overall, potato is rich in potassium but poor in sodium.

Of the carbohydrates comprising 65-80% of the dry weight of the tuber, important sugars are sucrose, glucose, fructose, galactose, melibiose, raffinose, stachyose, planteose, myinositol, maltotriose and manniinotriose. Cellulose is present in the ratio of 10-20% of nonstarch polysachharides.

References : Agarwal (2003), Ambasta (1986), Anonymous (1948-1976), Watt (1889-1896).

Solena amplexicaule Lamk.

Syn : *Melothria heterophylla* (Lour.) Cogn.;
Zehneria umbellata (Klein *ex* Willd.) Thw.
Z. umbellate (Seringe) Clarke. var. *nepalensis.*
Family : Cucurbitaceae.

Hindi and Regional Names:

Hindi : Tarali;
Beng. : Kudari;
Khasi. : Lushai, Soh-khia-phalang Zongaum pawng;
Mal. : Njerijanpuli.

Distribution : Common throughout India upto 2,100m.
Reproductive Cycle : April –January (Next year).
Part Used : Roots.
Folk Use : Roots eaten boiled.
Reference : Bennet *et al.* (1991).

Stachys palustris Linn.

Family : Lamiaceae.

English and Hindi Names:

Eng. : Marshy Wound Wort;
Hindi : Baggibuti.

Distribution : Trans-N.W. Himalaya (Kashmir).
Part Used : Roots.
Folk Use : Roots edible alongwith its shoots.
Reference : Agarwal (2003).

Stachys sieboldii Miq.

Syn : *S. tubifera* Naud.

Family : Lamiaceae.

English Names:

Eng. : Chinese Artichoke, Japanese Potato.

Distribution : Himalaya.

Part Used : Tubers.

Folk Use : Tubers eaten.

Reference : Ambasta (1986).

Stephania wightii A. Dunn.

Family : Menispermaceae.

Distribution : Kerala.

Part Used : Root Tubers.

Folk Use : Root tubers consumed as vegetable by the 'Irular' tribe of Kerala.

References : Jain (1996), Radhakrishnan *et al.* (1996).

Sterculia urens Roxb.

Family : Sterculiaceae.

Regional Name :

Mar. : Kadai.

Distribution : Outer and Subhimalayan zone and in Peninsular India.

Part Used : Tender Roots.

Folk Uses : Tender roots cooked and eaten after mixing with either spices or sugar.

References : Arora & Pandey (1996). Gammie (1902).

Sterculia villosa Roxb.

Family : Sterculiaceae.

English, Hindi and Regional Names:

Eng. : Udal;
Hindi : Udar;
Ass. : Odol;
Bihar. : Ganjher;
Garo. : Ubak;

Mar. : Sarda, Sardal;
Ori. : Kodalo.

Distribution : Throughout the greater parts of India upto 1,000m. Andaman Islands.

Part Used : Roots.

Folk Use : Roots of this tree eaten by the tribals living in the hills of 'Netarhat' plateau of Bihar.

References : Bennet *et al.* (1991), Manna & Samadar (1984).

Tacca integrifolia Ker-Gewl.

Syn : *T. aspera* Roxb.
Family : Taccaceae.

Hindi, Sanskrit and Regional Names:

Hindi : Varahi-kanda;
Sans. : Varahakanda;
Guj. : Dukarkanda;
Kan. : Handigadde;
Mar. : Dukarkand, Varahkand.

Distribution : E and N.E. Himalaya (Assam, Arunanchal Pradesh).

Part Used : Tubers.

Folk Uses : Sweet tubers considered nourishing tonic, and also eaten as vegetable.

References : Agarwal (2003), Ambasta (1986), Anonymous (1948-1976), Arora & Pandey (1996).

Tacca laevis Roxb.

Family : Taccaceae.

Distribution : N.E.Himalaya. Peninsular India.

Part Used : Tubers.

Folk Uses : Tubers sweet and edible. Also the extracted starch used as a substitute for the arrowroot (*Manihot esculenta*).

References : Ambasta (1986), Anonymous (1948-1976), Arora & Pandey (1996).

Tacca leontopetaloides (Linn.) Kuntze.

Syn : *T. pinnatifida* Forst. & Forst.f.

Family : Taccaceae.

English, Hindi, Sanskrit and Regional Names:

Eng. : Fiji Arrowroot, Tahiti Arrowroot;
Hindi : Surna;
Sans. : Surna;
Bomb. : Deva Kanda, Diva;
Kan. : Kaddu-churangle;
Mal. : Kattuchena;
Mar. : Karachunnie;
Tam. : Karachenie-kandam;
Tel. : Pedda -kanda -jadda.

Distribution : Wild in N.E. Himalaya, Bengal, C. India and W. Peninsula.

Reproductive Cycle : August- September.

Part Used : Tubers.

Folk Uses : Boiled tubers eaten and yield starch called 'Tahiti' which is used as a substitute for true arrowroot (*Manihot esculenta*). This starch has excellent culinary properties and used for making bread, cakes, porridges, sweetmeat, etc., and also recommended as a food for invalids.

Active Constituents : The starchy flour comprises of fibres, 3.0; nitrogenous substances, 0.01, starch, 76.00 and water, 18.0 %. However, its bitterness is due to taccalin ($C_{18}H_{26}O_7$, 0.003%).

References : Agarwal (2003), Ambasta (1986), Anonymous (1948-1976), Arora & Pandey (1996), Bennet *et al.* (1991), Krishna *et al.* (1999), Vartak (1981), Watt (1889-1896).

Tacca pinnatifida J.R. & G. Forsk.

Syn : *T. sitorea* Rumph.; *T. pinnatifolia* Gaertn.

Family : Taccaceae.

Distribution : Hilly tracts of C. India.

Part Used : Tuberous Roots.

Folk Uses : Large tuberous roots yield considerable amount of nutritious 'fecula' resembling to arrowroot starch and used especially by the natives of 'Travancore'.

References : Vartak (1981), Watt (1889-1896).

Remarks : Bitterness of tubers are removed by frequent boiling and washing. It is also said to improve its flavour.

Taraxacum officinale Wiber ex Wigg.

Syn : *T.densleonis*

Family : Asteraceae.

English, Hindi and Regional Names:

Eng. : The Dandlion;
Hindi : Bararu, Dulal, Kanphul;
Kan. : Kaddu -seventhi;
Ladk. : Rasuk.

Distribution : Throughuot Temperate Himalaya. Hilly regions of S.India.

Reproductive Cycle : April-October.

Part Used : Roots.

Folk Uses : Roots employed as a substitute for 'Chikory' for mixing with coffee. Also eaten as salad or boiled for use as vegetable.

References : Agarwal (2003), Anonymous (1948-1976), Arora & Pandey (1996), Rana (2004), Sinha (1996), Watt (1889-1896).

Telosma cordata (Burm. f.) Merrill.

Family : Asclepiadaceae.

Distribution : Entire Himalaya.

Part Used : Fleshy roots.

Folk Use : Roots used in the preparation of sweetmeat.

Reference : Arora & Pandey (1996).

Theriophonum indicum (Dalz.) Engl.

Family : Araceae.

Part Used : Roots

Folk Uses : Roots eaten at the time of food scarcity.

Reference : Arora & Pandey (1996).

Tigridia pavonia Ker-Gawl.

Family : Iridaceae.

English Name :

Eng. : Tiger or Shell Flower.

Distribution : N.E. and Tropical Himalaya.

Part Used : Bulbs.

Folk Use : Starchy bulbs have chestnut-like flavour and are eaten after roasting.

References : Agarwal (2003), Ambasta (1986), Anonymous (1948-1976).

Triglochin maritima Linn.

Family : Juncaginaceae.

Distribution : Trans-N.W. Himalaya (J.&K.): 3,500-4,800 m.

Part Used : Tuberous Corms.

Folk Use : Tuberous corms edible.

References : Arora & Pandey (1996), Samant *et al.* (2001).

Triglochin palustris Linn.

Family : Juncaginaceae.

Distribution : Trans-N.W. Himalaya (J.&K.): 3,000-4,000 m.

Part Used : Tubers.

Folk Use : Tubers edible.

References : Arora & Pandey (1996), Samant *et al.* (2001).

Tulipa gesneriana Linn.

Family : Liliaceae.

English Name :

Eng. : Common Tulip, Garden Tulip, Late Tulip.

Distribution : Cultivated in Indian Gardens.

Part Used : Bulbs.

Folk Use : Bulbs eaten at the time of food scarcity.

References : Agarwal (2003), Ambasta (1986), Anonymous (1948-1976).

Tulipa stellata Hook.

Family : Liliaceae.

Regional Names :

Pub. : Bhumphor, Chamuni, Jal Kukar Paduna, Piperi.

Distribution : W. Himalaya (Punjab, Shivalik Hills). Kumaon and Garhwal: 1,700 -2,300 m.

Part Used : Bulbs.

Folk Uses : Bulbs frequently eaten by natives and also sold in the village market.

References : Ambasta (1986), Arora & Pandey, (1996), Rana (2004), Samant *et al.* (2001), Watt (1889-1896).

Typha angustata Bory & Chaub.

Syn : *T. australis* Schum. & Thonn.;
T. elephantina Roxb.non Grah.nec. Schimp. *ex* Roxb.

Family : Typhaceae.

English and Hindi Names:

Eng. : Lesser Indian Reed Mace ;
Hindi : Patera.

Distribution : Marshy regions of India.

Part Used : Rhizomes.

Folk Use : Rhizomes eaten as vegetable.

References : Agarwal (2003), Ambasta (1986), Arora & Pandey (1996), Raju (2000), Rana (2004), Rana *et al.* (2005), Samant *et al.* (2001), Watt (1889-1896).

Typha angustifolia Linn.

Syn : *T. elephantina* Roxb.

Family : Typhaceae.

English, Hindi and Regional Names:

Eng.	:	Elephant Grass, Lesser cat tail, The Reed Mace;
Hindi	:	Pater;
Beng.	:	Hogla;
Pub.	:	Boj, Bori, Gond Luck, Pan Patira, Pityara.

Distribution : Wetlands all over India.

Part Used : Roots.

Folk Use : Roots alongwith young shoots edible and taste like the roots of *Asparagus*.

Reference : Watt (1889-1896).

Typha elephantina Roxb. non Grah.

Family : Typhaceae.

English Name :

Eng.	:	Elephant Grass.

Distribution : Marshy regions throughout India.

Part Used : Rhizomes.

Folk Use : Rhizomes edible and used as vegetable.

References : Agarwal (2003), Arora & Pandey (1996), Rana (2004).

Typha latifolia Linn.

Family : Typhaceae.

English and Hindi Names :

Eng.	:	Elephant Grass;
Hindi	:	Mothitrina.

Distribution : Wetlands of Himalaya and S. India.

Part Used : Rhizomes.

Folk Use : Rhizomes constitute regular diet of some tribals.

References : Raju (2000), Watt (1889-1896).

Typha laxmanni Lepech.

Family : Typhaceae.

English Name :

Eng. : Scened Flag.

Distribution : Marshy regions of India.
Part Used : Rhizomes.
Folk Use : Rhizomes edible and used as vegetable.
References : Agarwal (2003), Arora & Pandey (1996), Rana *et al.* (2005).

Typhonium bulbiferum Dalz. DC.

Family : Typhaceae.

Distribution : Malabar and Konkan coast of Kerala.
Part Used : Bulbs.
Folk Use : Bulbs alongwith leaves eaten boiled.
References : Ambasta (1986), Watt (1889-1896).

Typhonium diverticum Wall. *ex* Schott.

Family : Araceae.

Distribution : C. Himalaya: 1,800-2,500 m.
Part Used : Tuberous Roots.
Folk Use : Tuberous roots edible.
References : Arora & Pandey (1996), Samant *et al.* (2001).

Typhonium trilobatum Schott. DC.

Syn : *Arum orixense* Roxb.; *A. trilobatum* Linn.; *T. orixense* Schott.; *T. roxburghii* Saunders.; *T. simense* Zeigler.

Family : Typhaceae.

Hindi and Regional Names:

Hindi : **Gher-kochu;**
Ass. : **Samakaru;**
Beng. : **Gher-Kochu;**
Mal. : **Karuna -kizhanga;**
Tam. : **Karunai-kizhangu;**
Tel. : **Duradda- kanda -gadda.**

Distribution : Himalaya.

Folk Use : Cooked tubers eaten alongwith tender leaves.

Active Constituents : Various constituents of root tubers on dry matter basis are ash, 0.6-1.5; carbohydrates, 26.0; fats, 0.1; fibres, 1.0; minerals, 1.6 (Ca, 35; P, 20; Na, 9.0; K, 237 mg/100g,); moisture, 69.9; proteins, 1.4%; niacin, 0.7; riboflavin, carotene, 78; thiamine, 0.07; folic acid, 17.5mg/100g; I, 0.8 ppm; F, 3.7 ppm; and Cl as chlorides, 22.85 mg/100g.

References : Agarwal (2003), Ambasta (1986), Anonymous (1948-1976), Arora & Pandey (1996), Watt (1889-1896)

Remarks : Tubers acrid and be consumed by boiling and roasting.

Utleria salicifolia Bedd.

Family : Asclepiadaceae.

Regional Name :

Mal. : Mahalikighangu.

Distribution : S. India.

Part Used : Root Tubers.

Folk Use : Root tubers considered good for pickling by 'Irular' and 'Muthuvan' tribes of Kerala.

Reference : Radhakrishanan *et al.* (1996).

Urtica tuberosa Roxb.

Family : Urticaceae.

Distribution : S. India.

Part Used : Tuberous Roots.

Folk Use : Tuberous roots eaten boiled.

Reference : Shortt (1887-1888).

Vanda tessellata (Roxb.) D.Don.

Family : Orchidaceae.

Distribution : E.Himalaya (Assam and Arunanchal Pradesh) upto 1,000 m.

Part Used : Tender Roots.
Folk Use : Tender roots edible.
Reference : Samant *et al.* (2001).

Vigna adenantha (Meyer) Marechal.

Syn : *Phaseolus adenanthus* Meyer.
Family : Papilionaceae.

Regional Names :

Beng. : Banbarbati;
Bomb. : Kullonda;
Tam. : Kattupa-yaru;
Tel. : Karela sana.

Distribution : Throughout the plains of India.
Part Used : Tuberous Roots.
Folk Use : Tuberous roots eaten by locals in Chota Nagpur at the time of scarcity.
Reference : Bennet *et al.* (1991).

Vigna capenesis Chand & Arora.

Family : Papilionaceae.

Distribution : Cultivated all over India.
Part Used : Tuberous Roots.
Folk Use : Tuberous roots edible.
Active Constituents : Tuber comprises of albuminoids, 3.4, carbohydrates, 18.9; fats, 0.2 and moisture, 76.5 %.
Reference : Anonymous (1948-1976).

Vigna trilobata Linn. Verdcourt

Syn : *Dolochus trilobatus* Linn.; *Phaseolus trilobatus* (linn.) Schreb.; *P. trilobus* Sensu Ait. & Auct.; *Dolichus trilobus* Linn.
Family : Papilionaceae.

Hindi and Sanskrit Names:

Hindi : Mugam, Rakhalkalai, Trianguli;
Sans. : Mudgaparni.

Distribution : Himalaya.

Part Used : Roots.

Folk Use : Roots yield a kind of arrowroot-like starch and eaten.

Reference : Ambasta (1986).

Vigna vexillata (Benth.) A. Rich.

Family : Papilionaceae.

Hindi, Sanskrit and Regional Names:

Hindi : Halunda;
Mah. : Halunda;
Meg. : Jermei-soh-lang-tor;
Sant. : Ghangra.

Distribution : N.E. Himalaya. Throughout the Tropical parts of India upto 2,000 m.

Reproductive Cycle : September- October.

Part Used : Tuberous Roots.

Folk Uses : Small carrot-sized tubers eaten raw, boiled or roasted by N.E. tribals and in Maharashtra. Tuberous fusiform rootstocks considered superior to sweet potato in flavour and nutrients

References : Ambasta (1986), Arora (1981), Arora & Pandey, (1996), Bennet *et al.* (1991), Samant *et al.* (2001), Watt (1889-1896).

Vitis lanata Roxb.

Family : Vitaceae.

Distribution : Himalaya and Peninsular hilly tracts.

Part Used : Roots.

Folk Use : Roots edible.

Reference : Arora & Pandey (1996).

Xyris pauciflora Willd.

Family : Xyridaceae.

Hindi Name :

Hindi : Baluyanjuri.

Distribution : Marshy places of C. and E. India.
Part Used : Bulbs.
Folk Uses : Bulbs edible.
References : Ambasta (1986), Arora & Pandey (1996).

Yucca gloriosa Linn.

Family : Liliaceae.

English Name :

Eng. : Adam's needle.

Distribution : Grown in gardens and also found wild in lower foothills of Himalaya.
Part Used : Roots.
Folk Use : Roots edible.
References : Ambasta (1986), Watt (1889-1896).

Zingiber cassumunar Roxb.

Syn : *Amomum zingiber* Linn.; *Cassumunar longifolia* Wall.; *Z. missionis* Wall.

Family : Zingiberaceae.

Hindi, Sanskrit and Regional Names:

Hindi : Banada;
Sans. : Vanadaraka;
Beng. : Banadrak;
Mar. : Nissana.

Distribution : Throughout India.
Reproductive Cycle : July- December.
Part Used : Rhizomes.
Folk Use : Rhizomes having camphoraceous odour are used as a condiment and spices for flavouring food preparations. These are sold in markets as true cassumnar.
Active Constituents : Compounds responsible for flavour and pungency are zingiberene, curcumine, camphor, linalool and zingiberol, etc.
References : Agarwal (2003), Ambasta (1986), Anonymous (1948-1976), Watt (1889-1896).

Zingiber montanum (Koen.) Thel.

Family : Zingiberaceae.

Distribution : Tropical Himalaya upto 1,200 m.

Part Used : Rhizomes.

Folk Uses : Rhizomes used as spice and condiment for flavouring food preparations and also as a substitute for true ginger.

References : Mitrre (1981), Tripathi & Goyal (2001).

Zingiber officinale Rosc. (Pl. 9A, B)

Syn : *Amomum zingiber* Linn.; *Z. missionis* Wall.; *Cassumnar longifolia* Wall.

Family : Zingiberaceae.

English, Hindi, Sanskrit and Regional Names:

Eng. : East Indian Ginger, Ginger, Zinger ;
Hindi : Adrak, Adhruka;
Sans. : Adrakam, Anupama, Chandrakhya, Nagara, Sunthi, Vara;
Beng. : Ada, Adrok;
Kan. : Naisunmthi, Vansunthi.

Distribution : Widely cultivated in Tropical India & Himalaya upto 1,200 m.

Reproductive Cycle : August-October.

Part Used : Rhizomes.

Folk Use : Fresh and dried rhizomes highly esteemed as spice and condiment for its characteristic odour and warm chilly pungent taste and used mainly in culinary preparations, pickles, canned ginger and ginger cocktails, and also for flavouring pies, syrups and soups, puddings, cookies, pastries and biscuits, etc. It is also used as an excellent preservative.

Active Constituents : Composition of ginger samples dried at 55-60 °C for 8 hours shows ash, 5.1-9.3 (6.6); carbohydrates (starch), 40.4-59.0(53.0); fats, 0.3; crude fibres, 4.8-9.8 (7.2); minerals, 1.0 %, moisture, 8.5-16.5 (10.9);

proteins (crude), 10.3-15.0 (12.4); alcoholic extract, 3.6-9.3 (6.0); acetone extract, 3.9-9.3 (6.5)); water extract, 14.4-25.8 (19.6); and volatile oils, 1.0-2.7 (1.8) %. However, analysis of green ginger reveals the occurrence of carbohydrates, 12.3; fats, 0.9; fibres, 2.4; minerals, 1.2 (Ca, 20; P, 60; Fe, 2.6 mg/100g, I, 0.82 ppm; F, 2.0 ppm); proteins, 1.2 %; vitamins as carotene, 40ìg/100g; niacin, 0.60; riboflavin, 0.03; thiamine, 6.06, vit. C, 6.0 mg /100g and moisture, 80.9%. Major constituents of oil responsible for its pungency, odour and flavour are zingiberene, gerinial, ar-curcumene, cinole, camphene, zerombone and gingirole.

References : Agarwal (2003), Ambasta (1986), Anonymous (1999), Arora & Pandey, (1996), Aswal (2001), Jain (1995), Kirtikar & Basu (1984), Mehtani & Ingle (1999), Rastogi & Mehrotra (1991, 1995), Samant *et al.* (2001), Sharma & Singh (2001) Tripathi & Goyal (2001), Watt (1889-1896).

Zingiber zerumbet Rosc. ex Sm.

Syn : *Amomum spurium* Gmel.; *A. zerumbet* Linn.; *Z. spurium* Kon.

Family : Zingiberaceae.

Hindi, Sanskrit and Regional Names:

Hindi : Mahabari -bach, Narkachur;
Sans. : Ahava, Avanti, Kapuraharidra, Viranum;
Mal. : Kattingi;
Tel. : Karralamu.

Distribution : Throughout India.

Part Used : Rhizomes.

Folk Uses : Rhizomes used as spice and condiments, and also cooked or added raw in vegetables.

Active Constituents : Oils responsible for flavour are zerombone, humulene, cucurmine, etc.

References : Ambasta (1986), Arora (1981), Arora & Pandey, (1996), Aswal (2003), Bennet *et al.* (1991), Borthakur (1996), Haridasan *et al.* (1990), Islam (1996), Samant *et al.* (2001), Watt (1889-1896).

Zizania aquatica Linn.

Family : Poaceae.

English and Hindi Names:

Eng. : Water Rice;

Hindi : Jangli Dhan.

Part Used : Rhizomes.

Folk Use : Rhizome edible.

Reference : Raju (2000).

Zizania caducifolia Hand.-Mazz.

Syn : *Limnochloa caducifolia* Turz. apud Trin.; *Z. latifolia* Turz. *ex* Stapf.

Family : Poaceae.

Hindi Name :

Hindi : Pani ka Bans.

Distribution : Wetlands of N.E. Himalaya.

Part Used : Rhizomes.

Folk Use : Rhizomes eaten during food scarcity.

References : Agarwal (2003), Ambasta (1986), Singh *et al.* (2001b)

Epilogue

The information collected in the present database on edible roots and underground stems of ethnic India has brought to light the use of 401 plant species belonging to 186 genera under 67 families as perspective food plants which could be utilized as alternative food or in times of scarcity and famine. Following conclusions emerge from the statistical analysis of the gathered data:

1. Of the 401 edible species, 8 spp (6 genera: *Angiopteris, Blechnum, Isoetes, Marsilea, Nephrolepis, Pteridium*) belong to Pteridophytes; 1 sp (1 genus: *Cycas*) to Gymnosperms; 197 spp (117 genera: *Abelmoschus, Abutilon, Acanthopanax, Aconitum, Ajuga, Allophylus, Althaea, Ampelocissus, Angelica, Anredera, Antigonon, Aponogeton, Armoracia, Arnebia, Arcticum, Aspidopterys, Borassus, Beta, Boussingaultia, Brachystelma, Brasenia, Brassica, Bunium, Bupleurum, Butea, Calamus, Careya, Caralluma, Carum, Ceropegia, Chaerophyllum, Cichorium, Cinnamomum, Cirsium, Clematis, Coptis, Codonopsis, Coleus, Corydalis, Costus, Cotoneaster, Crambe, Daucus, Decalepis, Dicentra, Dracocephalum, Echinopsis, Elephantopus, Eriosema, Erycibe, Erythropsis, Euryale, Ferula, Flemingia, Geranium, Glycyrrhiza, Helianthus, Hemidesmus, Heracleum, Houttuynia, Ipomoea, Kedrostis, Leea, Leptadaenia, Limnanthemum, Litsea, Manihot, Marrubium, Marsdenia, Meconopsis, Melothria, Merremia, Mirabilis, Momordica, Moringa, Mucuna, Nelumbo, Nymphaea, Nymphoides, Onosma, Orthosiphon, Oxalis, Pachyrhizus, Paeonia, Passiflora, Pastinaca, Pentatropis, Persea, Peucedanum, Phaseolus, Pleurospermum, Polygonum, Portulaca, Potentilla, Pouzolzia, Pueraria, Raphanus, Rheum, Rumex, Sechium, Sedum, Silybum, Sium, Solanum, Solena, Stachys, Stephania, Sterculia, Tacca, Taraxacum, Telosma, Triglochin, Utleria, Urtica, Vigna, Vitis, Xyris*) to Dicots (Magnoliopsida); and 195 spp (62 genera: *Acorus, Alisma, Allium, Alocasia, Alpinia, Amischophacelus, Amomum, Amorphophallus, Arisaema, Asparagus, Asphodelus, Chlorophytum, Clinogyne, Colocasia, Commelina, Crinum, Curculigo, Curcuma, Cyanotis, Cyperus, Dactylorhiza, Didymospermum, Dioscorea, Dracontium, Eleocharis, Eria, Etlingera, Eulophia, Gagea, Gastrochilus, Gonatanthus, Habenaria, Hedychium, Hypoxis, Kaempferia, Lasia, Leucojum, Lilium, Maranta, Paris, Phragmites, Phyllostachys, Plesmonium, Polygonatum, Potamogeton, Raphidophora, Remustia, Sagittaria, Sauromatum, Scirpus, Scleria, Sinobambusa, Smilax, Theriophonum, Tigridia, Tulipa, Typha, Typhonium, Vanda, Yucca, Zingiber, Zizania*) to Monocots (Liliopsida). The dicotyledonous taxa predominate in their use.
2. From the distribution view-point, most of the edible roots and underground stems plant spp having wider distribution throughout India predominate in usage, followed by 88 spp in N.W. and W. Himalaya, 65 spp in N.E. and E. Himalaya, 60 spp

in Peninsular India (Ghats, Coastal Plains, Deccan Plateau), 53 spp in Central Highlands and Plains of India and 41 spp in Entire Himalaya (Table IV; Fig. 1). It appears that the roots and underground stems are relished by all Indian tribals located in various physiographical regions.

3. For edible purposes, the most frequently used families for roots and underground stems are Liliaceae (45 spp, 11 genera), Dioscoreaceae (36 spp, 1 genus), Araceae (31 spp, 14 genera), Zingiberaceae (30 spp, 8 genera), Paplionaceae (20 spp, 10 genera), Apiaceae (21 spp, 12 genera), Asclepiadaceae (17 spp, 10 genera), Polygonaceae (13 spp, 4 genera), Brassicaceae (12 spp, 4genera), Cyperaceae (11 spp, 4 genera), Convolvulaceae (11 spp, 3 genera), Asteraceae (10 spp, 8 genera) Nymphaeaceae (9 spp, 4 genera), Orchidaceae (8 spp, 5 genera), Lamiaceae (8spp, 5 genera), Typhaceae (8 spp, 2 genera), Cucurbitaceae (5 spp, 5 genera), Ranunculaceae (5 spp, 4 genera), Poaceae (5 spp, 4 genera), Potamogetonaceae (5 spp, 1 genus), Malvaceae (4 spp, 3 genera), Alismataceae (4 spp, 2 genera), Gentianaceae (4 spp, 2 genera), Taccaceae (4 spp,1 genus), etc. (Appendix II)

4. Five important dicotyledonous families which predominate for edible roots and underground stems are Papilionaceae, Apiaceae, Asclepiadaceae, Polygonaceae and Brassicaceae whereas for monocotyledonous taxa these are: Liliaceae, Dioscoreaceae, Araceae, Zingiberaceae and Cyperaceae (Appendix II).

5. Amongst the dicotyledonous taxa, the predominant edible genera are those of *Polygonum* (10 spp), *Ipomoea* (7 spp), *Brassica* (7 spp), *Ceropegia* (6 spp), *Nymphaea* (5 spp), *Carum* (4 spp), *Pueraria* (4 spp), *Tacca* (4 spp), *Vigna* (4 spp), *Aponogeton* (3 spp), *Bupleurum* (3 spp), *Coleus* (3 spp), *Ferula* (3 spp), *Flemingia* (3 spp), *Merremia* (3 spp), *Nymphoides* (3 spp), *Portulaca* (3 spp) and *Abelmoschus, Aconitum, Angelica, Caralluma, Chaerophyllum, Cirsium, Codonopsis, Crambe, Phaseolus, Potentilla, Raphanus, Sedum, Silybum, Solanum, Stachys, Sterculia, Triglochin* etc. with 2 species each. Likewise, these for monocots are: *Dioscorea* (36 spp), *Allium* (21 spp), *Curcuma* (12 spp), *Arisaema* (9 spp), *Alpinia* (6 spp), *Asparagus* (6 spp), *Amorphophallus* (5 spp), *Cyperus* (5 spp), *Potamogeton* (5 spp), *Typha* (5 spp), *Colocasia* (4 spp), Polygonatum (4 spp), *Scirpus* (4 spp), *Zingiber* (4 spp), *Alocasia* (3 spp), *Asphodelus* (3 spp), *Habenaria* (3 spp), *Kaempferia* (3 spp), *Sagittaria* (3 spp), *Smilax* (3 spp), *Typhonium* (3 spp), *Chlorophytum* (2 spp), *Tulipa* (2 spp), *Zizania* (2 spp), etc. (Appendix III).

Table - II

Nutritional Components of Some Edible Roots and underground stems of Ethnic India with moisture (%)}*

Name of the Plant	Moisture	Carbohydrates	proteins	Fats	Ash	Fibres	Minerals	Vitamins
Allium cepa	86.3	11.6	1.2	<0.1	-	-	Ca, P	A, B, C
Allium porrum	78.9	17.2	1.8	0.1	-	-	Ca, P,	A, B, C
Allium sativum	62.2	29.0	6.3	0.1	-	-	Ca, P,	A, B, C
Amorphophallus campanulatus	78.4	18.4	1.2	0.1	-	0.7	Ca, P, Fe	-
Asparagus officinalis	93.0	3.2	2.2	0.2	-	-	Ca, P, Fe	A, B, C
Asparagus racemosus	68.5	28.2	1.2	0.3	-	-	Ca, P, Cu	-
Beta vulgaris	76.6	20.4	1.1	-	8.5	1.1	Ca, P, Fe	-
Boussingaultia baselloides	85.0	10.5	2.3	-	13.5	0.4	-	-
Brassica rapa	91.0	7.6	0.5	0.2	-	-	Ca, P, Fe	A, B, C
Cichorium intybus	77.0	-	-	<0.6	-	9.0	K	-
Coles parviflorus	77.6	19.7	1.3	0.1	-	0.4	-	-
Colocasia esculenta	73.1	22.1	3.0	0.1	-	-	Ca, P,	A, B, C
Dioscorea wallichi	79.6	18.9	1.2	0.1	-	-	-	-
Helianthus tuberosus	71.6-84.2	8.0-18.0	0.9-3.25	0.29	2.5	0.3-3.0	Ca, P, Fe, Al, Cl, I, K	A, B, C
Ipomoea batatus	68.5	31.0	1.2	0.3	1.2	-	Ca, P, Fe	A, B, C
Manihot esculenta	69.4	38.7	0.7	0.2	-	0.6-2.5	Ca, P, Fe, Mg, Cl, I	A, B, C
Maranta arundinacea	63.4	27.8	1.6	0.2	0.9	3.9	-	-
Nelumbo nucifera	83.80	9.66	2.70	0.8	1.2	0.6-0.25	Ca, P, Fe	-
Nephrolepis biserrata	53.95	36.14	5.87	1.06	-	-	-	-
Oxalis martiana	74.4	0.7	0.7	-	-	-	-	-
Pachyrrhizus erousus	82.38	10.74	1.47	0.09	0.5	0.64	-	B, C
Pastinaca sativa	72.4	23.2	1.3	0.3	1.2	-	Ca, P,Fe	A, B, C
Pueraria lobata	68.6	27.8	2.1	0.1	-	0.7	Ca, P, Fe	-
Raphanus sativus	94.4	3.4	0.7	0.1	-	0.8	Ca, P, Fe	-
Sagittaria sagittaefolia	69.5	23.3	5.0	0.2	-	-	Ca, P, Fe	-
Sagittaria trifolia	69.5	24.7	5.0	0.3	-	-	-	-
Solanum tuberosum	75.4-77.0	18.9-21.4	1.8-2.3	0.2	-	-	Ca, P, Fe, Mg, Cl, I	-
Typhonium trilobatum	69.9	26.0	1.4	0.1	-	-	Na, K	A, B, C
Vigna capnesis	76.5	18.9	3.4	0.2	-	-	-	-

Table - III

Nutritional Components of Some Edible Roots and underground stems of Ethnic India {on dry matter basis (%)}*

Name of the Plant	Carbohydrates	Albuminoids	Fats	Fibres	Moisture	Organic Acid	Ash	Minerals
Arisaema concinnum	65.9	7.7	1.7	8.9	8.5	-	-	Ca, 1.2; P, 0.2
Arisaema speciosum	76.0	7.7	1.6	6.1	7.6	-	-	Ca, 1.1; P, 0.5
Curcuma angustifolia	26.00-46.00	30.73	0.2	8.42	13.13	-	4.46	-
Curcuma caesia	18.75	22.94	0.2	25.25	9.76	10.0	7.75	-
Curcuma caulina	10.00-18.00	-	-	-	-	-	-	-
Curcuma domestica	26.00	6.3	5.1	-	13.10	-	-	3.5
Curcuma zedoria	82.0	0.5	0.2	-	13.10	-	-	-
Cyperus bulbosus	82.46	2.87	0.58	1.51	11.6	-	1.58	-
Cyperus esculentus	24.79	5.21	-	5.83	14.15	25.82	-	1.48
Dioscorea alata	71.67-85.2	7.96-15.18	0.9	2.19-6.12	-	-	6.0	-
D. bulbifera	75.1-81.39	7.36-13.31	9.5	3.28-9.64	-	-	5.0	P, 0.45
D. deltoidea	71.29	10.82	1.72	6.92	-	-	9.65	P, 0.92
D. esculenta	77.29	12.02	1.02	5.52	-	-	6.65	P, 0.94
D. glabru	77.79-78.23	9.73-10.13	1.35	3.92-5.0	-	-	-	P, 0.58
D. hamiltonni	85.80	8.30	0.7	1.52	-	-	4.35	P, 0.55
D. hispida	81.85-81.89	7.20-9.12	1.15	3.28-6.28	-	-	-	P, 0.52
D. opositifolia	68.54	14.70	1.75	6.55	-	-	-	P, 0.71
D. pentaphylla	71.67-85.02	7.96-15.8	0.9	2.19-6.12	-	-	6.0	P, 0.85
D. punber	78.42-81.34	11.44-12.45	084	2.94-3.3	-	-	3.5	P, 0.55
Eleochairs dulsis	7.47	8.12	0.7	3.70	4.34	-	-	-
Flemingia tuberosa	40.42	13.4	-	12.16	-	1.64	3.44	-
Hedychium coronarium	30.02	1.65	0.33	29.6	-	9.2	2.3	-
Nymphaea alba	46.00	6.4	7.8	10.0	-	-	10.8	-
Nymphaea lotus	6.78	9.3	9.3	5.81	10.94	-	6.32	-
Nymphoides stellata	67.49	14.15	-	6.25	4.20	-	-	-
Pteridium aquilinium	51.0	9.5	9.5	20.0	-	-	8.3	-
Pueraria tuberosa	64.6	10.9	10.9	28.8	10.0	-	-	-
Scirpus grossus	76.18	4.12	4.12	6.80	8.40	-	3.40	-
S. lacustris	10.14	6.10	6.10	40.0	-	-	7.84	-
S. tuberosus	10.3	10.3	10.3	71.0	-	-	7.84	-
Zingiber officinale	53.0	12.4	0.3	7.2	1.0	12.0	0.9	Ca, 1.2

Table IV

Distribution Patterns of Edible Roots underground stems and Rhizomes of Ethnic India.

Distribution	No. of species
Entire India or Scattered Distribution	94
N.W. and W. Himalaya	88
N. E. and E. Himalaya	65
Peninsular India (Ghats, Coastal Plains, Deccan Plateau)	60
Central Highlands and Plains of India	53
Entire Himalaya	41

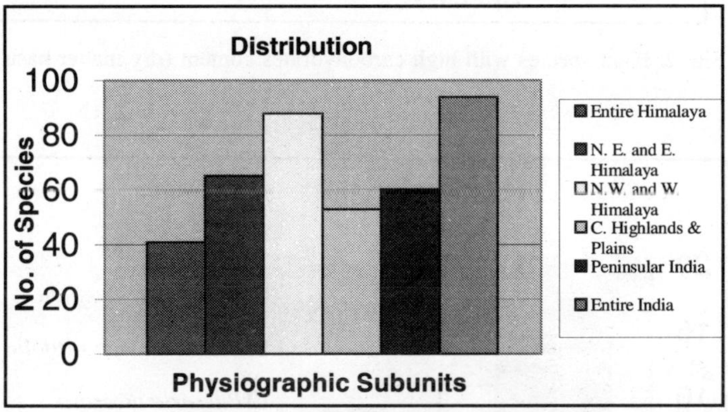

Fig. 1. Histogram showing distribution patterns of edible roots and underground stems of India.

6. Most of the plant species are eaten raw or after boiling and also cooked as vegetable and even pickled, thus are playing a major role in day to day life of the people of India. The tribals owe their health, vitality and longevity to these wild edibles.

7. Tribals are also well-versed with the art of processing and preparing the harvested edible part of most of the plant species as these are thoroughly washed before eating to eliminate the toxic content.

8. Data on nutrient analysis reveals that *Dioscorea hamiltonni, Cyperus bulbosus, Eleocharis dulsis, Scirpus grossus and Pueraria tuberosa* are rich in *carbohydrates* (Fig. 2); *Nymphoides stellata, Dioscorea oppositifolia, Pueraria tuberosa, Pteridium aquilinum* and *Nymphaea lotus* are rich in proteins (Fig. 3) whereas *Arisaema* concinnum, Curcuma domestica and Dioscorea bulbifera are rich in fat content (Fig. 4). For the sustainable utilization of the important species, mass scale propagation in the nurseries and agricultural fields needs to be encouraged.

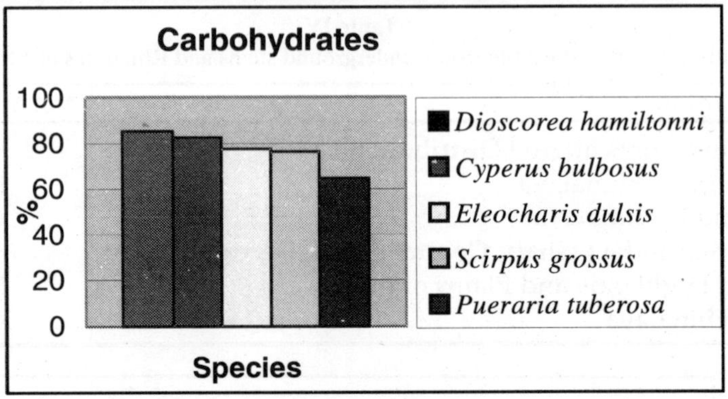

Fig. 2. Plant species with high carbohydrates content (dry matter basis)

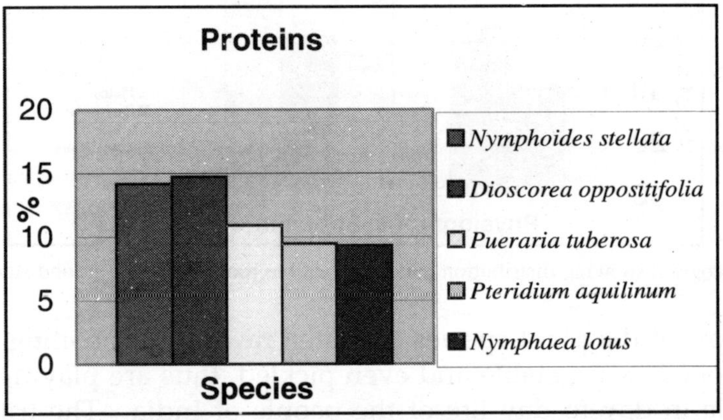

Fig. 3. Plant species with high proteins content (dry matter basis)

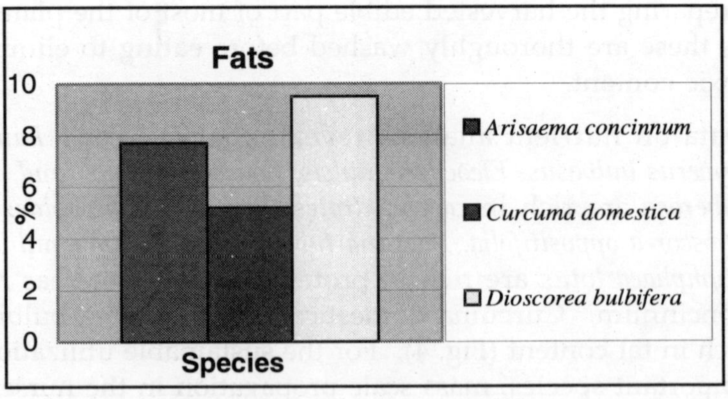

Fig. 4. Plant species with high fats content (dry matter basis)

9. The generated data will hopefully be utilized in developing effective strategies for their wider consumption. Also, it would be useful to horticulturists and others by enabling production of improved varieties of edibles plants for meeting the food demand of growing population as well as to overcome the sufferings due to the deficiency of nutritional constituents.

LITERATURE CITED

Abraham, Z. 1981. Ethnobotany of Todas, the Kotas and Irulas of Nilgiris, 308-320. In: Jain S.K. (*ed.*) *Glimpses of Indian Ethnobotany*. Oxford & IBH Pub. Co., New Delhi.

Agrawal, V.S. & Malhotra, S.K. 1970.® Unreported economic plants of India. Indian Mus. Bull. *5(1)*: 101-105.

Agarwal, V.S. 1986. *Economic Plants of India*. Kailash Prakashan, Calcutta.

Agarwal, V.S. 2003. *Directory of Indian Economic Plants*. Bishan Singh, Mahendra Pal Singh, Dehradun.

Ambasta, S.P. (*ed.*) 1986. *The Useful Plants of India*. C.S.I.R., New Delhi.

Anonymous, 1948-1976. *The Wealth of India, Raw Materials*. Vol. I-XI. P.I.D., C.S.I.R., New Delhi.

Anonymous, 1958. Plants of economic importance, Lucknow district. Bot. Garden Bull. 9.

Anonymous, 1986a. *Economic Plants of India*. P.I.D., C.S.I.R., New Delhi.

Anonymous, 1986b. *The Useful Plants of India*. P.I.D., C.S.I.R., New Delhi.

Anonymous, 1994. *All India Co-ordinated Research Project on Ethnobiology* (*Annual Report*). Department of Environment, New Delhi.

Anonymous, 1999. *The Ayurvedic Pharmacopoeia of India*, Part 1. Vol. II. Govt. of India- Ministry of Health & Family Welfare, India.

Ansari, A.A. 1997. Less known wild edible plants of Shevory and Kolly hills of South India, 243-247. In: Jain, S.K. (*ed.*) *Contributions to Indian Ethnobotany*, 3rd ed. Sci. Pub., Jodhpur.

Arora, R.K. 1981. Native food plants of northeastern tribals, 91-106. In: Jain, S.K. (*ed.*) *Glimpses of Indian Ethnobotany*. Oxford & IBH Pub. Co., New Delhi, Bombay, Calcutta.

Arora, R.K. 1985a. Life support species of Himalayan region, 42-45. In: Paroda, R.S., Kapoor, P., Arora, R. K. & Bhagmal. *Life Support Plants Species, Diversity and Conservation*, N.B.P.G.R., New Delhi.

Arora, R.K. 1985b. *Genetic Resources of Less Known Cultivated Food Plants*. N.B.P.G.R., New Delhi.

Arora, R.K. 1991. Native food plants of tribals of northeast India, 137-152. In: Jain, S.K. (*ed.*) *Contributions to Indian Ethnobotany*. Sci. Pub., Jodhpur.

Arora, R.K. & Nayer, E.R. 1984. *Wild Relatives of Crop Plants of India*. Sci. Monograph. Vol.-7. N.B.P.G.R., New Delhi.

Arora, R.K. & Pandey, A. 1996. *Wild Edible Plants of India. Diversity, Conservation and Use*. N.B.P.G.R., New Delhi.

Asolkar, L.V., Kakkar, K.K. & Chakre, O.J. 1992. *Second Supplement to Glossary of Indian Medicinal Plants with Active Principle, Part-1 (A-K)*. C.S.I.R., New Delhi.

Aswal, B.S. 2001. Zingiberaceae of western Himalaya, 101-118. In: Pandey, P.C. & Samant, S. S. (*eds.*) *Plant Diversity of the Himalaya*. Gyanodaya Prakashan, Nainital.

Atal, C. K., Sharma, B.M. & Bhatia, A. J. K. 1980. Search of emergency food plants through wild flora of Jammu and Kashmir State, Sundarbani area. Indian Forester *106* : 211-219.

Atkinson, E.T. 1982. *Economic Botany of the Himalayan Region*. Cosmo Pub., New Delhi.

Badhwar, R.K. & Fernandez, R.R. 1964. *Edible Wild Plants of Himalaya*. New Delhi. India.

Badhwar, R.L., Rao, P.S. & Sethi, H. 1964. *Some Useful Aromatic Plants*. F.R.I., Dehradun.

Bajpayee, K.K. & Dixit, G. 1996. Ethnobotanical studies on foodstuff of tribals of Tarie-region, Uttar Pradesh. J. Econ. Taxon.. Bot. *12*: 128-132.

Bal, S.N. 1942. Useful plants of Mayurbhanj State in Orissa. Rec. Bot. Sur. Ind. 6: 1-112.

Barrau, J. 1959. The Sago Palms and other food plants of marsh dweller in South Pacific island. Econ. Bot. *13*: 151-162.

Bennet, S.S.R.. 1987. *Name Changes in Flowering Plants of India and Adjacent Regions*. Triseas Pub., Dehradun, India.

Bennet, S.S.R., Biswas, S., Chandra,V., Singh, P. J., Chandra, S. & Dayal, R. 1991. *Food From Forest*. F.R.I., I.C.F.R.E., Dehradun.

Bhandari, M.M. 1974. *Famine Foods in Rajasthan Desert*. Sci. Pub., Jodhpur.

Bhandari, M.M. 1979. Famine food of Rajasthan desert. Econ. Bot. *28*: 73-81.

Bhargava, K.S. 1959. Usual and supplementary food plants of Kumaon. J. Bomb. Nat. History Soc. *56*: 26-31.

Bhargava, N. 1981. Plants in folklife and folklore in Andaman and Nicobar Island, In: Jain, S.K. (*ed.*) *Glimpses of Indian Ethnobotany*. Oxford and IBH Pub. Co., New Delhi.

Bhattacharjee, S.K. 1998. *Handbook of Medicinal Plants*. Pointer Pub., New Delhi.

Biswas, S. & Bhuyan, T.C. 1983. On the identity of some food plants of Garo hills, Meghalaya. Indian J. Forestry *6(3)*: 208-213.

Bonde, S.D., Agte, V., & Kulkarni, D.K. 1990. Nutritional composition of the fruits and the Down Palms (Hyphaene) from the west coast of India. Principles *34*: 21-23.

Borthakur, S.K. 1996. Certain plants in folklore and folklife of the Karbis (Mikirs) of Assam, 170-181. In: Jain, S.K. (*ed.*) *Glimpses of Indian Ethnobotany*. Oxford & IBH Pub. Co., New Delhi.

Borthakur, S.K. 1996. Wild edible plants in the market of Assam, India. An ethnobotanical investigation, 31-34. In: Jain, S.K. (*ed.*) *Ethnobotany in Human Welfare*. Deep Pub., New Delhi.

Burkill, I. H. 1935. *Dictionary of the Economic Products of the Malay Peninsula*. Oxford Univ. Press, England.

Cameron, L.C.R. 1917. *The Wild Foods of Great Britain*. London.

Castetter, E.F. 1935. Uncultivated nature plants used as a source of food. Ethnobiological studies in the American south-west- I. Univ. New Mexico Bull. 266.

Chandra, V. 1997. *Edible Plants of Forestry Origin*. I.C.F.R.E., Dehradun.

Chaudhary, B. 1967. *Vegetables*. National Book Trust, New Delhi.

Cheng, T. 1965. Utilization of wild plants in communist China. Econ. Bot. *19* : 3-15.

Chevallier, A. 1996. *The Encyclopedia of Medicinal Plants*. Dorling Kindersley.

Chopra, R.N., Nayer, S.L. & Chopra, I.C. 1956. *Glossary of Indian Medicinal Plants*. C.S.I.R., New Delhi.

Chopra, R.N., Chopra, I.C. & Verma, B.S. 1968. *Suppliment to Glossary of Indian Medicinal Plants*. C.S.I.R., New Delhi.

Clute, W.N. 1943. *Useful Plants of the World*. Indiana.

Dager, J.C. & Dager, H.S. 1987. Some useful pteridophytes of Andaman & Nicobar Islands. J.Econ. Tax. Bot. *9*: 317-323.

Dalzell, J.M. 1937. *The Useful Plants of Tropical West Africa*. London.

Das, H.S., Panda, P.C. & Patnaik, S.N. 1996. Traditional uses of wetland plants of eastern Orissa. Jour. Econ. Tax. Bot. *12*: 306 -313.

Das, N. & Chattopadhyay, R.N. 2001. Nutritional evaluation of edible non timber forest produce- A case study in South West Bengal. Rural Development Centre, IIT Khargpur (West Bengal). Indian Forester *127*: 1232-1238.

Dastur, J. F.1951. *Useful Plants of India and Pakistan*. D.B.Taporowala & Sons, Bombay.

Datar, R. & Vartak, V.D. 1975. Enumeration of wild edible plants from Karnala bird sanctuary , Maharashtra State. Biovigayanum *1*: 23-129.

Deb, B.B. 1975. Economic plants of Tripura State- Edible plants, food grains, vegetable and fruits. Indian Forester *101*: 399-405.

Dhar, U., Rawal, R.S. & Samant, S.S. 1998. Wild plant resources and use pattern in Askot wild life sanctuary of Kumaon in the Indian Himalaya. Manifestation of commonness and rarity, 128-142. In: Kotwal, P.C. & Banerjee (*ed.*) *Biodiversity Conservation in Managed Forest and Protected Area*. Agrobotanica, Bikaner.

Dutt. U.C. 1993. *Materia Medica of Hindus*. Mittel Pub., New Delhi.

Fernald, M.L. & Kensy, A.C. 1943. *Edible Wild Plants of Eastern North America*. New York.

Gammie, G. A. 1902. A note on plants used for food during famines and seasons of scarcity in the Bombay Presidency, India. Bot. Surv. Records 2 : 171-196.

Gaur, R.D. 1977. Wild edible fruits of Garhwal hills. The Himalaya *1(1)*: 66-70.

Gaur, R.D. 1997. Wild edible plants of Garhwal hills the Himalaya. Man & Environment *1*: 66-70.

Gaur, R.D. & Semwal, J. K. 1983. Some little known wild edible of Garhwal Himalaya. Man & Environment *7* : 161-165.

Girach, R.D., Amminiddin & Ahmed, I. 1988. Observation on wild edible plants from triblal pocket of Orissa. Plant Sci. Res. *10(1)*: 16-25.

Gopalan, C., Ranasastri, B.V. & Balasubramanyam, S.C. 1971. *Nutritive Value of Indian Foods*. Nat. Inst. Nut., Indian Council of Medical Research, Hyderabad.

Goud, P., Prasad, S. & Pullaiah, T. 1999. Ethnobotany of Kurnool District (Andhra Pradesh). Some wild plants used as food. Jour. Econ. Tax. Bot. *12*: 224-227.

Greenish, H.G. 1999. *Materia Medica.* Sci. Pub., Jodhpur.

Gupta, R. K. 1962. Some unusual and interesting food plants of the Garhwal Himalayas. J. d Agr. Trop. et de Bot. Appl. *9* : 532-535.

Gupta, R. K. & Kanodia, K.C. 1968. Plants used during scarcity and famine periods in the dry regions of India. J. d Agr. Trop. et de Bot. Appl. *15*: 265-285.

Hajra, P. K. & Chakraborty, P. 1981. A survey of wild plants sold in Lal market of Gangtok, India. J. Econ..Taxon.. Bot. *4* : 217-220.

Harborne, J.B. & Baxtar, H. 2001. *Chemical Dictionary of Economic Plants.* John Wiley & Sons.

Haridasan, K., Bhuyan, L.R. & Deori, M.L. 1990. Wild edible Plants of Arunanchal Pradsh. Arunanchal Forest News. *8 (192)* : 1-8.

Hill, J. 1939. *Wild Foods of Britain.* London.

Hooker, J. D. 1872-1897. *The Flora of British India,* Vol. I-VII. London.

Hosagouder, V.B. & Henery, A.N. 1996. Ethnobotany of Kadars, Malasars and Muthuvan of the Annamalai in Coimbtore district, Tamil Nadu. India. Jour. Econ.Tax.Bot. *12*: 260-267.

Irvine, F. R. 1952. Supplementry and emergency food plants of west Africa. Econ. Bot. *6*: 23-40.

Islam, M. 1996. Ethnobotany of certain underground parts of N.E. region, India, 338-343. In: Maheshwari, J. K. (*ed.*). *Ethnobotany in South Asia.* Sci. Pub., Jodhpur.

Jain, S.K. 1963. Wild plant-foods of tribals of Baster. Proc. Nat. Instt. India *30B*: 56-80.

Jain, S.K. 1964. The role of botanists in folk-lore research. Folk-lore 5: 145-150.

Jain, S.K. 1967. Ethnobotany-its scope and study. Indian Museum Bull. *2*: 39-43.

Jain, S.K. 1968. *Medicinal Plants.* National Book Trust, India.

Jain, S.K. 1981. Observation on ethnobotany of tribals of Central India, 193-198. In: Jain S.K. (*ed.*) *Glimpses of Indian Ethnobotany.* Oxford & IBH Pub. Co., New Delhi.

Jain, S.K. 1986. Ethnobotany. Interdiscipl. Sci. Rev. *II*: 285-292.

Jain, S.K. 1987. *A Manual of Ethnobotany.* Sci. Pub., Jodhpur.

Jain, S.K. 1995. Ethnobotanical diversity in zingibers of India. Ethnobotany *1*: 83-88.

Jain, S.K. 1996. Wild food plants of the tribals of Bastar (Madhya Pradesh). Proc. Nat. Inst. Sci. India *30b* : 56-80.

Jain, S.K. 2002. *Bibliography of Indian Ethnobotany*. Sci. Publ., Jodhpur.

Jain, S.K. & De, J.N. 1964. Some less known plants foods among the tribals of Purulia, West Bengal. Science and Culture. *30*: 285.

Jain, S.K. & Hajra, P.K. & Shanpru, R. 1977. Survey of edible plants in Bazaar of Meghalaya. Bull. Meghalaya. Sci. Society 2: 29-34.

Jain, S.K., Mudgal, V., Banerjee, D.K., Guha, A., Pal, D.C. & Das, D. 1984. *Bibliography of Ethnobotany*. 157. B.S.I., Howrah.

Jain, S.K. & Sinha, B.K. 1987. Ethnobotanical aspects of life support species-some emergency and supplementary foods of aborginals in India, 173-180. In: *Life Support Plants Species, Diversity and Conservation.* N.B.P.G.R., New Delhi.

Jamir, N.S. 1995. Studies on wild edible fruits in Nagaland State, India. J. Non-wood For. Prod. 2 *(192)*: 79-82.

Janaki Ammal, E.K. 1955. Introduction to the subsistence economy of India, 324-335. In: William, L.T. (*ed.*) *Men's Role in Changing Face of the Earth*. Univ. of Chicago Press, Chikago.

Joseph, J. & Kharkongar, P. 1996. A preliminary ethnobotanical survey in Khasia and Jantia Hills, 25-29. In: Jain S.K. (*ed.*) *Glimpses of Indian Ethnobotany*. Oxford & IBH Pub. Co., New Delhi.

Kapur, S.K., Nanda, S. & Srivastava, T.N. 1996. Traditionally important medicinal and aromatic plants of Bhaderwah hills–Jammu provimnce III, 62-69. In: Maheshwari, J.K. (*ed.*) *Ethnobotany in South Asia*. Sci. Pub., Jodhpur.

Kaul, A.K., Karihaloo, J.L., & Hamal, L.A. 1982. Wild edible plants of Kashmir-some less known vegetable substitute and beverages. Bull. Bot. Surv. of India *24*. 67-69.

Kaul, M.K., Bhatia, A.K. & Atal C.K. 1985. Ethnobotanical studies in north-west and trans-north west Himalaya. Contribution to wild food plants of Laddakh. J. Econ. Tax. Bot. *6*: 523-587.

Kaul, M.K. & Singh, S. 1985. Wild edible plants of ethnobotanical significance . Sci. & Culture *51*: 381- 388.

King, G. 1869. Famine foods of Marvar. Proc. Asiat. Soc. Bengal *38*: 116-122.

Kirtikar, K. R. & Basu, B.D. 1984. *Indian Medicinal Plants*. Vol. 1-4. Lalit Mohan Basu, Allahabad, India.

Koelz, W.N. 1979. Notes on the ethnobotany of Lahul, a province of the Punjab. Quar. J. Crude Drug Res. *17*: 1-56.

Kolesnikov, B.P. 1943. *Edible Wild Plants*. Moscow, Russia.

Krishna. P., Rajagopal, T., Kant, Y. & Badrinath, K.V.S. 1999. Food plants of Konda Reddies of Rampa Agency, East Godavari district, Andhra Pradesh –a case study. Ethnobotany 11: 92-96.

Krishnamurthy, S. (*ed.*) 1953. *Horticulture and Economic Plants of Nilgiris*. Govt. Bot. Gdn., Ootty, India.

Kshirsagar, S.R., Parabia, M.H. & Reddy, M.N. 2003. Ethnobotany of coastal area of Gujarat. Ethnobotany *15*: 60-63.

Kulkarni, D.K., Agte, V.V. & Kumbojkar, M.S. 2003. Leafy vegetable consumed by Mahadeokoli tribe in Western Maharashtra with their nutritional potential. Ethnobotany *15*: 34-38.

Kulkarni, D.K. & Kumbhojkar, M.S. 1994. Palatable fodder grasses from Pachgaon Parvati area in Pune District. J. Econ. Tax. Bot. *19*: 529-532.

Kumar, R. V. & Pullaiah, T. 2003. A survey of wild edible plants of Chenchu tribes of Prakasm district (Andhra Pradesh). Vistas in Ethnobotany *1*: 22-31.

Kunkal, G. 1984. *Plants for Human Consumption*. Koeltz Sci., Books, W. Germany.

Lal, B., Vats, S. K., Singh, R. D. & Gupta, A. K. 1994. Plants used as ethnomedicine and supplementary food by Gaddies of Himachal Pradesh, India, 384-387. In: Jain, S.K. (*ed.*) *Ethnobotany in Human Welfare*. Deep Pub., New Delhi.

Lal, G., Tondon, G.L. & Pruthi, J. S. 1952: Utilization of tender Bamboo shoots as an article of food. Chemical Age *6*: 110-114.

Lalramnghinglova, J.H. 2002. Ethnobotanical study on the edible plants of Mizoram. Ethnobotany *14*: 23-33.

Lancaster, S.P. 1964. The food plants of India. Bull. *et*. Bot. Garden. 99.

Maheshwari, J.K. 1987. Tribal ecosystem-an overview. Tribal Res. Bull. 9: 1-14.

Maheshwari, P. & Singh, U. 1965. *Dictionary of Economic Plants of India*. I.C.A.R., New Delhi.

Malhotra, C. L., Balodi, B. S. & Singh, S. 1985. Additional notes on edible plants India. J. Econ. Tax. Bot. 2: 481-482.

Mandal, S.K. & Basu, S.K. 1996. Ethnobotanical studies among some tribals of Nilgiri district, Tamilnadu. Jour. Econ. Tax. Bot. *12*: 268-271.

Maniyam, S. & Ramachandran, V.S. 1990. Wild edibles of Coimbatore district, India. J.Econ. Tax. Bot. ***14(3)***.

Manna, M. K. & Samadar, U. P. 1984. On the occurence of the useful plants of Netarhat plateau of Palamou district, Bihar. J. Econ. Tax. Bot. *8*: 539-546.

Medsger, O.P. 1943. *Edible Wild Plants*. New York.

Mehra, K.L., Kanodia, K.C. & Srivastava, R.N. 1975. Folk uses of plants for adornment in India. Econ. Bot. *29*: 39-46.

Mehtani, S. & Ingle, P.K. 1999. *Plants Food Flavours*. Nat. Inst. Sci. Communication, New Delhi.

Meninger, E.A. 1977. *Edible Nuts of the World*. Hortic. Books Inc., Stuart.

Mittre, V. 1981. Wild plants in Indian folklife- a historical perspective, 35-53. In: Jain, S.K. (*ed.*) *Contributions to Indian Ethnobotany*. Sci. Pub., Jodhpur.

Nadkarni, K.M. 1914. *Indian Plants and Drugs with Their Medicinal Properties and Uses*. Nortan & Co., San Thome, Madras.

Nadkarni, K.M. 1954. *Indian Materia Medica*. Vol. 1-2, 3rd edn. Popular Book Depot, Bombay.

Nathawat, G.S. & Deshpande, B.D. 1960. Plants of economic importance from Rajasthan. Proc. Raj. Acad. Sci. *7* : 38-48.

Nayer, M.P., Ramamurthy, K. & Agarwal, V.V. 1989-1994. *Economic Plants of India, Vol. I & II*. B.S.I., Calcutta.

Nazarudeen, A., Seensi, S., Koshy, K.C. & Pushpangadan, P. 1996. Folk plants of food, medicine, adornment and repellent used by the Pariyar community in north Kerala, 299-305. In: Maheshvari, J.K. (*ed.*) *Ethnobotany of South Asia*. Sci. Pub., Jodhpur.

Negi, K.S. 1988. Some little known wild edible plants of U.P. hills. J. Econ. Tax. Bot. *2(2)*: 345-360.

Negi, K.S. & Gaur, R.D. 1994. A contribution to the edible wild fruits of Uttar Pradesh hills. Bull. Bot. Surv. India *33*: 235-266.

Negi, K.S. & Pant, K.C. 1992. Less known wild species of *Allium* Linn. (Amaryllidaceae) from mountain regions of India. J. Econ.Taxon Bot. *46 (1)*: 112-116.

Negi, K.S., Pant, K.C. & Gaur, R.D. 1990. Sea-buckthorn (*Hippophae*). A pickle plant from central Himalaya. Acta. Bot. Indica *18*: 274-275.

Nilegaonkar, S., Vartak, V.D. & Chitre, R.G. 1985. Nutritional values of some wild food plants from Pune and neighbouring districts. Maharashra State. Part-I, J.Econ. Tax. Bot. *6*: 629-635.

Pal, D.C. & Banerjee, D.K. 1971. Some less known food plants among the tribals of Andhra Pradesh and Orissa State. Bull. Bot. Surv. India. *13*: 221-223.

Pandey, P.C. & Pangety, Y.P.S. 1987. Studies in ethnobotany-I. On some less known edible and economic ferns of Kumaon region of western Himalaya. J.Econ. Tax. Bot. *11*: 81-85.

Pandey, P.C. & Samant, S. S. 2001. *Plants Diversity of Himalaya.* Gyanodaya Prakashan, Nainital.

Pangtey, Y.P.S. 1980. Some wild edible fruits of Kumaon Hills, 350-363. In: Singh, J.S. *et al.* (*ed.*) *Science of Rural Development in Mountains.* Gyanodaya Prakashan, Nainital.

Pangtey, Y.P.S., Rawat, G.S. & Kalkoti, B.S. 1982. Usual and supplementary wild food plants of Kumaon. Himal. Res. & Dev. *1*: 35-40.

Parmar, C. & Kaushal, M.K. 1982. *Wild Fruits of the Sub-Himalayan Region.* Kalyani Publishers, Ludhiana, New Delhi.

Paroda, R.S., Kapoor P., Arora, R. K. & Bhagmal 1985. *Life Support Plants Species, Diversity and Conservation.* N.B.P.G.R., New Delhi.

Patole, S. N. & Jain A. K. 2002. Some wild edible plants of Panchmahri biosphere reserve (M.P.). Ethnobotany *14* : 48-51.

Porsild, A.E. 1937. *Edible Roots and Berries of Northern Canada.* Nat. Mas. Canada, Ottawa.

Portefied, M.M. 1951. Principal Chinese vegetable foods and food plants of china town market. Econ. Bot. *5*: 3-37.

Prasad, N.P. & Abraham, Z. 1984. Notes on ethnobotany of Lahul, a province of Punjab. Quart. J. Crude Drugs. Res. *17*: 1-56.

Pruthi, J.S. 1993. *Major Spices of India. Crop Management and Post Harvest Technology.* I.C.A.R., New Delhi.

Purohit, A.N. 2000. Medicinal plants. Need for upgrading technology for trading traditions, 49-75. In: Nautiyal, A. R., Nautiyal, M.C. & Purohit, A.N. (*eds.*) *Harvesting Herbs. Medicinal and Aromatic Plants. An Action Plan of Uttarakhand.* Bishan Singh, Mahendra Pal Singh, Dehradun.

Radhakrishnan, K., Pandurangan, A.G. & Pushpagadan, P. 1996. Ethnobotany of wild edible plants of Kerala, India, 41-51. In: Jain, S.K. (*ed.*) *Ethnobotany in Human Welfare.* Deep Publication, New Delhi.

Raichaudhary, H.N., Dam, S.N., Pal, D.C., Tarafdar, C.R. & Banarjee, K.K. 1977. Edible jungle leaves. Folklore *18*: 161-163.

Raichaudhary, H.N., Pal, D.C., Saha, N.C. & Roy, B. 1985. Some wild edible plants in Culcutta markets. Man & Life *11*: 45-58.

Raju, D.C.S. & Krishna, B. 1988. Wild edible plants of Sikkim. Proc. Nat. Sym. Environ. Protection & Hill Dev.: 38-44. Sikkim Sci. Soc., Sikkim.

Raju, D.C.S. & Krishna, B. 1990. Less known edible plants of Sikkim, 83-86. In: Purkayastha, R.P. (*ed.*) *Economic Plants & Microbes.* Today & Tomorrow's Printer & Pub., New Delhi.

Raju, R.A. 2000. *Wild Plants of Indian Subcontinent and their Economic Use*. C.B.S. Pub., New Delhi.

Rana, J.C. 2004. *Lesser Known Food Plants of Himachal Pradesh*. N.B.P.G.R., Regional Station, Phagli, Shimla (H.P.).

Rana, J.C., Sharma, B.D., Sood, S.K., & Lakhanpal, T.N. 2005. *Genetic Diversity of Himalaya*. Publication No. 7. Inst. Int. Himalayan Studies, H.P.U., Shimla (H.P.).

Rao, R.N. & Henery, A.N. (ed.) 1996. *Ethnobotany of Eastern Ghats in Andhra Pradesh*, India. B.S.I., Kolkatta.

Rao, R. M. 1914. *Flowering Plants of Travancore*. Govt. Press, Trivendrum.

Rao, V.M.K. & Shanpru, R. 1981. Some plants in the life of Garos of Meghalaya, 153-166. In: Jain, S.K. (*ed.*) *Glimpses of Indian Ethnobotany*. Oxford & IBH Pub. Co., New Delhi.

Rastogi, R. P. & Mehrotra, B. N. 1990. *Compendium of Indian Medicinal Plants*. Vol. I 1970-1979. C.D.R.I. Lucknow. P.I.D., New Delhi.

Rastogi, R. P. & Mehrotra, B.N. 1991. *Compendium of Indian Medicinal Plants*. Vol. II. 1970-1979. C.D.R.I. Lucknow. P.I.D., New Delhi.

Rastogi, R.P. & Mehrotra, B.N. 1993. *Compendium of Indian Medicinal Plants*. Vol. III. 1970-1979. C.D.R.I. Lucknow. P.I.D., New Delhi.

Rastogi, R.P. & Mehrotra, B.N. 1995. *Compendium of Indian Medicinal Plants*. Vol. IV. 1970-1979. C.D.R.I. Lucknow. P.I.D., New Delhi.

Raveendran, J.C. 1996. Some lesser known edible plants among the tribals of Kerala, India, 52-53. In: Jain, S. K. (*ed.*) *Ethnobotany in Human Welfare*. Deep Pub., New Delhi.

Roy, B. Halder, A.C. & Pal, D.C. 1998. *Plants for Human Consumption in India*. B.S.I., Calcutta.

Sahu, T.R. 1996. Life supporting promising food plants among aborginals of Bastar (M.P.), India, 26-30. In: Jain, S.K. (*ed.*) *Ethnobotany in Human Welfare*. Deep Pub., New Delhi.

Saini, D.C. 1996. Ethnobotany of Tharus of Basti district, U.P. Jour. Econ. Tax. Bot. *12*: 138-153.

Saklani, A. & Jain, S.K. 1994. *Cross Cultural Ethnobotany of North East India*. Deep Pub., New Delhi.

Samant, S.S. & Dhar, U. 1997. Diversity, endemism and economic potential of wild edible plants of Indian Himalaya. Int. J. Sustain. Dev. & World Ecol. 179-191.

Samant, S.S., Dhar, U., & Rawal, R.S. 2001. Diversity and distribution of wild edible plants of the Indian Himalaya, 421-482. In: Pandey, P.C. & Samant, S.S. (*eds.*) *Plant Diversity of Himalaya*. Gyanodaya Prakashan, Nainital.

Sangal, P.M. 1971. Food plants of tribals population of Andaman and Nicobar islands. Indian Forester *97(7)* : 648-650.

Sarin, Y.K. 1989. Practical application of ethnobotany. Economic development of backward peoples, 167-178. In: Jain, S.K. (*ed.*) *Methods and Approaches in Ethnobotany*. Soc. Ethnobot., C.D.R.I., Lucknow.

Saunders, C.F. 1934. *Useful Plants of United States and Canada*. New York, U.S.A.

Saxena, S.K. 1979. Plant foods of Western Rajasthan. Man & Environment *3*: 35-43.

Schulltes, R.E. 1960. Tapping over heritage of ethnobotanical lore. Econ. Bot. *8*: 224-227.

Shankar, V. 1985. Life support species in Indian Thar desert, 37-41. In: Paroda, R.S., Kapoor, P., Arora, R.K. & Bhagmal *Life Support Plants Species, Diversity and Conservation*. N.B.P.G.R., New Delhi.

Sharma, B.D. & Malhotra, S.K. 1984. A contribution to the ethnobotany of tribals areas in Maharashtra. Jour. Econ. Tax. Bot. *5* : 533-537.

Sharma, K.R. & Sood, M. 1997. *Important Medicinal Plants of Himachal Pradesh, Vol. I.* Directorate of Ext. Edu., Dr. Y.S.Parmar Univ. Hort. & For., Nauni-Solan. (H.P.).

Sharma, P. 1996. Emergency food in Kautilaya's Arthashashtra-An ancient Sanskrit Work, 20-21. In: Jain, S.K. (*ed.*). *Ethnobotany in Human Welfare*. Deep Publication, New Delhi.

Sharma, P.P. & Singh N.P. 2001. Ethnobotanical uses of some edible plants of Dadra, Nagar and Haveli and Daman (U.T.) Ethnobotany *13*: 121-125.

Shortt, J. 1887-1888. List of wild plants and vegetables used as food by people in famine times. Indian Forester *3*: 232-238.

Shukla, K.M.L., Khan, A.A., Khan, S. & Verma, A.K. 2001. An ethnobotanical survey of wild edible plants of district Bilaspur (M.P.), India. Adv. Plant Science *14* : 57-60.

Singh, B., Chaurasia, O.P. & Ballabh, B. 2001a. Edible wild plants of trans-Himalayan cold desert, 483-512. In: Pandey P.C. & Samant S.S. (*eds.*) *Plant Diversity of Himalaya*. Gyanodaya Prakashan, Nainital.

Singh, B., Chaurasia, O.P., & Jadhav, K.L. 1996. An ethnobotanical study of Indus valley (Laddakh). J. Econ. Taxon. Bot. *12*: 92-101.

Singh, B.H., Hynniewta, T.M. & Bora, P.J. 1999. An ethnobotanical note on wild edible plants of Tripura, India. Ethnobotany *11*: 26-28.

Singh, G.S. 1999. Ethnobotanical study of useful plants of Kullu District in north-western Himalaya, India. J. Econ. Tax. Bot. *23*: 185-198.

Singh, H.B. & Arora, R. K. 1978. *Edible Wild Plants of India.* (1st ed.) I.C.A.R., Pub. New Delhi, India.

Singh, H.B. & Viswanathan, M.V. 1996. Useful pteridophytes of India- A gift of nature to human beings. J. Econ. Tax. Bot. *12*: 24-36.

Singh, K.K. & Anand, P. 1996. Observation on ethnobotany of Kol tribe of Varanasi district, U.P. India. J. Econ.Tax.Bot. *12*: 133-137.

Singh, K.K. & Kumar, K. 2000. *Ethnobotanical Wisdom of Gaddie Tribe of Western Himalayas.* Bishan Singh, Mahender Pal Singh, Dehradun.

Singh, K.K., & Maheshwari, J. K. 1985. Observation on some wild food plants of the tribals of Varanasi district, Uttar Pradesh. Vanyajati *33*(2): 25-31.

Singh, N.I. & Singh, L.J. 1988. Preliminary ethnobotanical studies on wild edible plants in the market of Manipur -II. J.Econ. Tax. Bot. *12*: 113-119.

Singh, P.B. 1996. Wild edible plants of Mandi district in N.W. Himalaya, 22-25. In: Jain, S.K. (*ed.*) Ethnobotany in Human Welfare. Deep Pub., New Delhi.

Singh, P.K., Elangban, B., & Huidrom, B.K. 2001b. Wild edible aquatic plants of Manipur valley, India, 513-522. In: Pandey, P.C. & Samant, S.S. (*eds.*) *Plant Diversity of Himalaya.* Gyanodaya Prakashan, Nainital.

Singh, S.R. & Singh, N. I. 1985. A preliminary ethnobotanical study on wild edible plants in market of Manipur. J. Econ. Tax. Bot. *6*.: 699-703.

Singh, U., Wadhawani, A.M. & Johri, B.M. 1983. *A Dictionary of the Economic Plants of India.* I.C.A.R., New Delhi.

Singh, V. & Pandey R.P. 1998. *Ethnobotany of Rajasthan, India.* Sci. Pub., Dehradun.

Singh, V. & Singh, P. 1981. Edible wild plants of eastern Rajasthan. J. Econ. Tax. Bot. 2: 197-207.

Sinha, S.C. 1996. Wild edible plants of Manipur, India, 42-47. In: Jain, S.K. (*ed.*) Ethnobotany in Human Welfare. Deep Pub., New Delhi.

Sood, S.K., Nath, Ram & Kalia, D.C. 2001. *Ethnobotany of Cold Desert Tribes of Lahul-Spiti* (N.W. Himalaya). Deep Pub., New Delhi,

Sood, S.K. & Thakur, S. 2004. *Ethnobotany of Rewalsar Himalaya.* Deep Pub., New Delhi.

Subramanyam, K. 1964. *Aquatic Angiosperms.* C.S.I.R., New Delhi.

Sundararaj, D.D. & Balasubramanyam, G. 1955. *Guide to the Economic Plants of South India.* Amudha Nilayam Pvt. Ltd. (Repr. 1986).

Tanaka, T. 1976. *Tanaka's Cyclopedia of Edible Plants of the World.* Keigku Pub., Co. Tokyo, Japan.

Tripathi, S. & Goyal, A. K. 2001. Ethnobotanical Diversity of Zingiberaceae in N. E. India. Ethnobotany *13*: 67-79.

Uma, J. & Singh, V. 1987. A census of edible species of Diospyros L. in India. J.Econ. Tax. Bot. *10.*: 416-419.

Vartak, V.D. 1959a. Some edible wild plants from the hilly regions of Poona district, Bombay State. J. Bom. Nat. Hist. Soc. *56(1)* : 7-25.

Vartak, V.D. 1959b. Some edible wild plants from the hilly regions of Maharashtra & Goa, 265-270. In: Jain, S.K. (*ed.*) *Glimpses of Indian Ethnobotany*. Oxford & IBH Pub. Co. New Delhi.

Vartak, V.D. 1981. Observation on wild edible plants form hilly regions of Maharashtra and Goa, resume and future prospect, 261-271 In: *Glimpses of Indian Ethnobotany*. Oxford & IBH Pub. Co., New Delhi.

Vartak, V.D. & Kulkarni, D.K. 1987. Monsoon wild leafy vegetables from hilly regions of Pune and neighbouring districts, Maharashtra State. J. Econ. Tax. Bot. *11(2)*: 331-335.

Vartak, V.D. & Suryanaryanan, M.C. 1995. Enumeration of wild edible plants from Susala island, Mulshi reservoir, Pune district. J. Econ Tax. Bot. *19(3)*: 555-569.

Watt, G. 1889-1896. *Dictionary of Economic Products of India*. Vol. I-VI. London.

Wester, P.J. 1925. *The Food Plants of the Philippines*. Dept. Agri. & Nat. Res. Bur. Agri. Bull. 39, Philippines.

Yawalakar, K.S. 1969. *Vegetable Crops of India*. Agri. Horti. Pub. House, Nagpur.

Pl. 1 A-D. Edible Roots and Underground Stems of Ethnic India

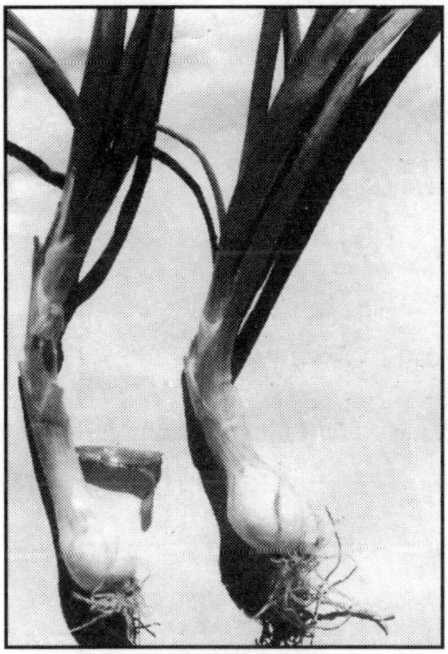

Fig. A. Entire plant of *Allium cepa* Linn.

Fig. B. Entire plant body of *Allium sativum* Linn

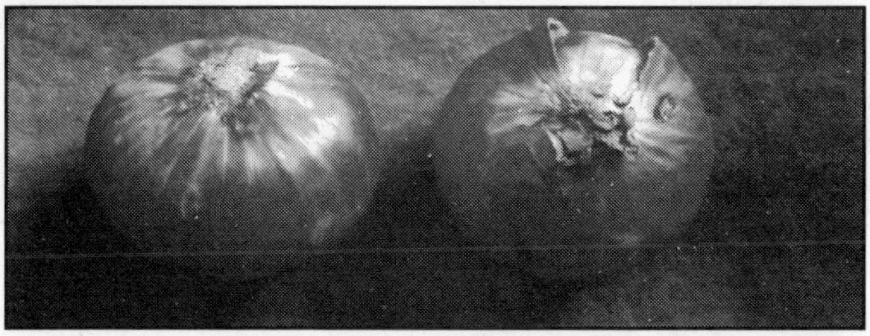

Fig. C. Bulbs of *Allium cepa* Linn.

Fig. D. Bulbs of *Allium sativum* Linn.

Pl. 2 A-C. Edible Roots and Underground Stems of Ethnic India

Fig. A. Large-sized corm of *Amorphophallus campanulatus* (Roxb.) Blume *ex* Decne.

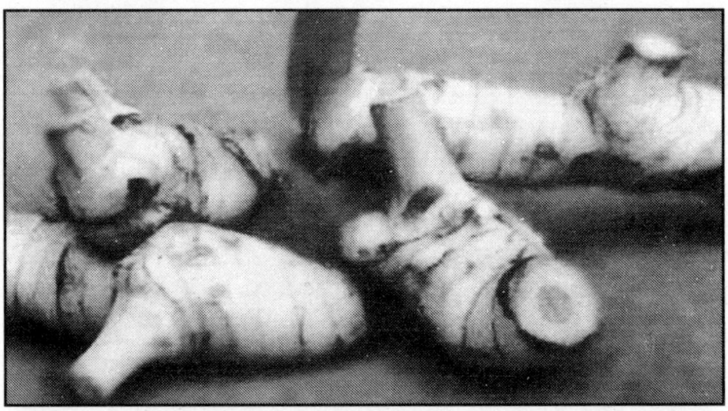

Fig. B. Rhizomes of *Angelica glauca* Edgew

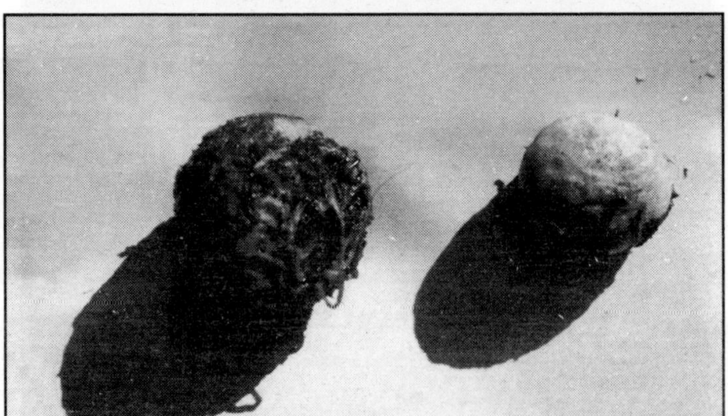

Fig. C. Tuberous corms *Arisaema sp.*

Pl. 3 A-D. Edible Roots and Underground Stems of Ethnic India

Fig. A. Entire plant of *Asparagus adscendens* Roxb. showing leaves and fasciculated roots.

Fig. B. Entire plant body of *Brassica rapa* Linn. showing napiform fleshy tap root.

Fig. C. Aerial plant parts of *Colocasia esculenta* (Linn.) Schott.

Fig. D. Lower plant body of *Colocasia esculenta* (Linn.) Schott. showing tubers and fibrous roots.

Pl. 4 A-C. Edible Roots and Underground Stems of Ethnic India

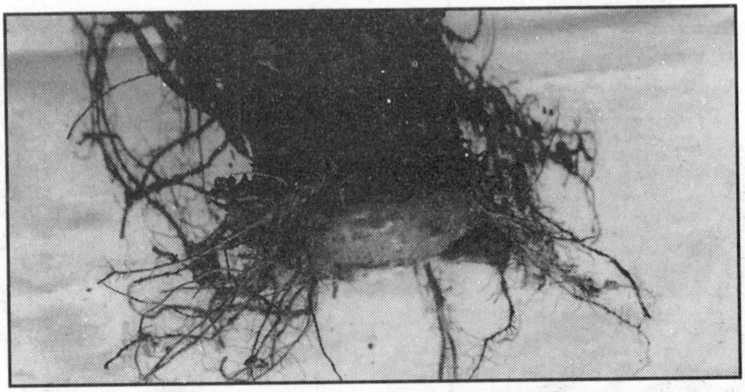

Fig. A. Enlarged tuber of *Colocasia* sp.

Fig. B. Aerial plant body of *Curcuma domestica* (Medic.) Velton.

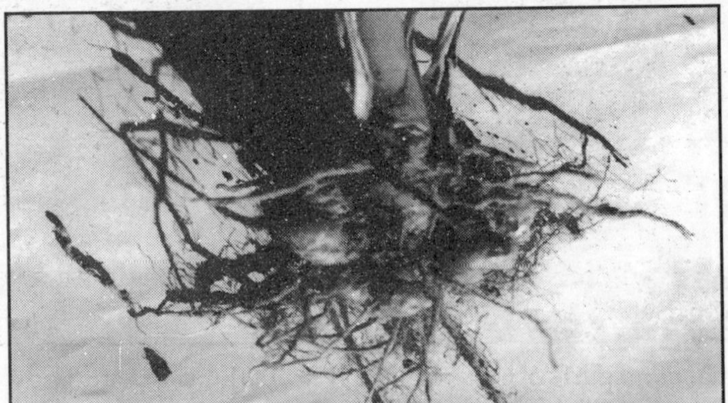

Fig. C. Rhizomes of *Curcuma domestica* (Medik.) Velton.

Pl. 5 A-C. Edible Roots and Underground Stems of Ethnic India

Fig. A. *Dioscorea belophylla* (Prain) Voight. *ex* Haines showing climbing habit and leaves.

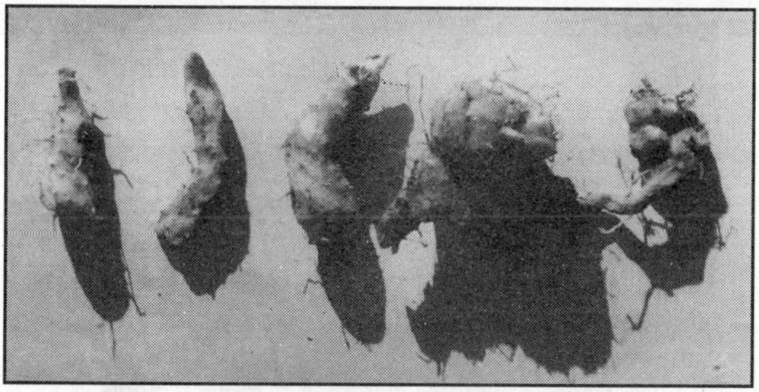

Fig. B. Tubers of *Dioscorea belophylla* (prain) Voight *ex* Hains.

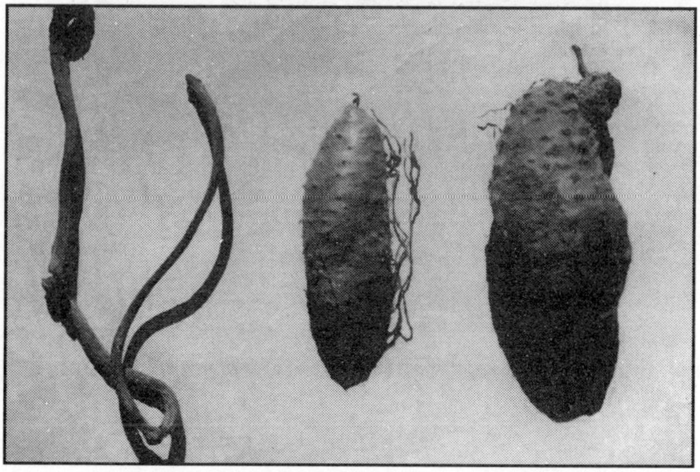

Fig. C. Tubers of *Dioscorea* sp.

Pl. 6 A-B. Edible Roots and Underground Stems of Ethnic India

Fig. A. Enlarged tubers of *Dioscorea hispida* Dennst.

Fig. B. Tuberous plant of *Dioscorea pentaphylla* Linn.

Pl. 7 A-C. Edible Roots and Underground Stems of Ethnic India

Fig. A. Rhizomes of *Hedychium spicatum* Buch. -Ham. *ex* Smith.

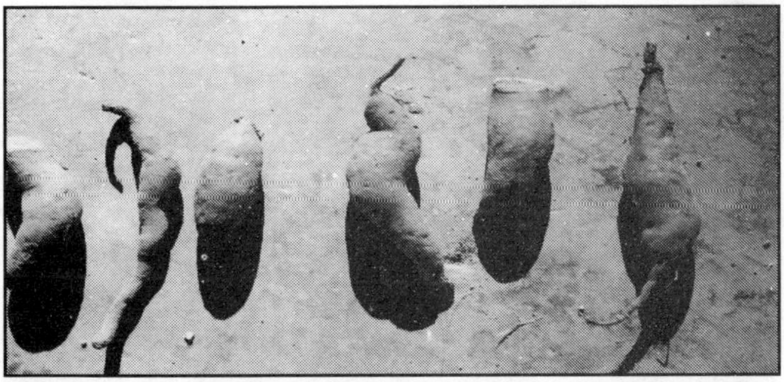

Fig. B. Tuber of *Ipomea batatus* Lamk.

Fig. C. Tubers of *Manihot esculenta* Crantz

Pl. 8 A-C. Edible Roots and Underground Stems of Ethnic India

Fig. A. Large-sized tubers of *Puerarta tuberosa* Rosb. *ex* Benth.

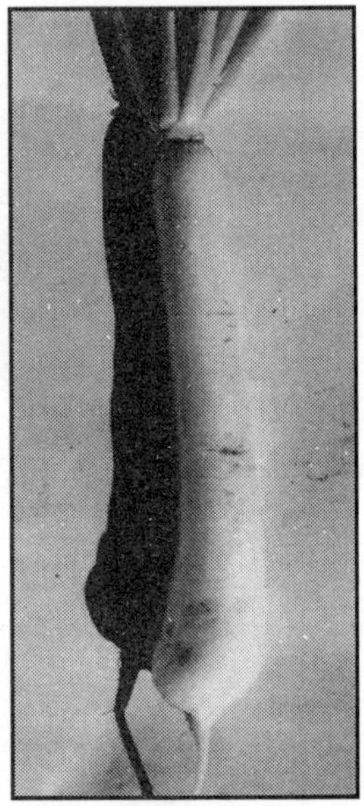

Fig. B. Fusiform fleshy tap root of *Raphanus sativus* Linn.

Fig. C. Stem tubers of *Solanum tuberosum* Linn.

Pl. 9 A-C. Edible Roots Underground Stems and Rhizomes of Ethnic India

Fig. A. Aerial plant body of *Zingiber officinale* Rosc.

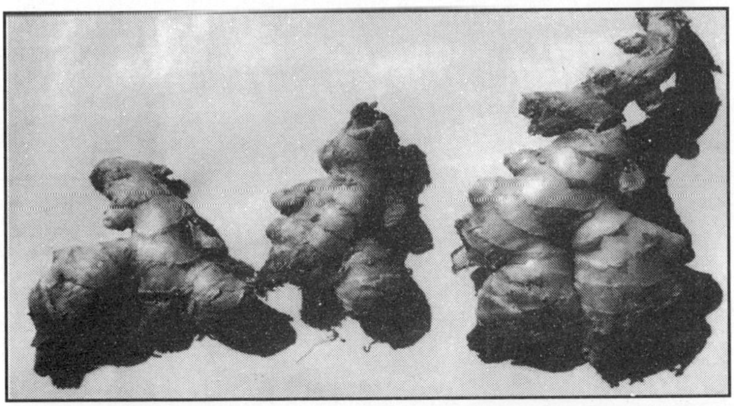

Fig. B. Rhizomes of *Zingiber officinale* Rosc.

Fig. C. Rhizomes of *Zingiber* sp.

APPENDIX - I

Index to Total Number of Edible Genera and Species of Roots and Underground Stems under Various Divisions of Plant Kingdom

Division	No. of Families	Total No. of Edible Genera/ Species	Total No. of Edible Species in a Genus
Pteridophytes	5	(6/8)	*Angiopteris (1), Blechnum (1), Isoetes(1), Marsilea (1), Nephrolepis (3), Pteridium (1).*
Gymnosperms	1	(1/1)	*Cycas (1)*
Dicots	46	(117/197)	*Abelmoschus (2), Abutilon(1), Acanthopanax (1), Aconitum (2), Ajuga (1), Allophylus (1), Althaea (1), Ampelocissus (1), Angelica (2), Anredera (1), Antigonon (1), Aponogeton (3), Armoracia (1), Arnebia (1), Arcticum (1), Aspidopterys (1), Borassus (1), Beta (1), Boussingaultia (1), Brachystelma (1), Brasenia (1), Brassica (7), Bunium (1), Bupleurum (3), Butea (1), Calamus (1), Careya (1),Caralluma (2), Carum (4), Ceropegia (6), Chaerophyllum (2), Cichorium (1), Cinnamomum (1), Cirsium (2), Clematis (1), Coptis (1), Codonopsis (2), Coleus (3), Corydalis (1), Costus (1), Cotoneaster (1), Crambe (2), Daucus (1), Decalepis (1), Dicentra (1), Dracocephalum (1), Echinopsis (1), Elephantopus (1), Eriosema (2), Erycibe (1), Erythropsis (1), Euryale (1), Ferula (3), Flemingia (3), Geranium (1), Glycyrrhiza (1), Helianthus (1), Hemidesmus (1), Heracleum (1), Houttuynia (1), Ipomoea (7), Kedrostis (1), Leea (1), Leptadaenia (2), Limnanthemum (1), Litsea (1), Manihot (2), Marrubium (1), Marsdenia (1),Meconopsis (2), Melothria (1), Merremia (3),Mirabilis (1),Momordica (1), Moringa (1), Mucuna (1), Nelumbo (2), Nymphaea (5), Nymphoides (3), Onosma (1), Orthosiphon (1),Oxalis(1), Pachyrhizus (2), Paeonia (1), Passiflora (1), Pastinaca (1), Pentatropis (1), Persea (1), Peucedanum (1), Phaseolus (2), Pleurospermum (1), Polygonum (10), Portulaca (3), Potentilla (2), Pouzolzia (1), Pueraria (4), Raphanus (2), Rheum (1), Rumex (1), Sechium (1), Sedum (2), Silybum (2), Sium (1), Solanum (2), Solena (1), Stachys (2), Stephania (1), Sterculia (2), Tacca (4), Taraxacum (1), Telosma (1),*

Division	No. of Families	Total No. of Edible Genera/ Species	Total No. of Edible Species in a Genus
			Triglochin (2), Utleria (1), Urtica (1), Vigna (4), Vitis (1),Xyris (1).
Monocots	15	(62/195)	*Acorus (1), Alisma (1), Allium (21), Alocasia (3), Alpinia (6), Amichophacelus (1), Amomum (1), Amorphophallus (5), Arisaema (9), Asparagus (6), Asphodelus (3), Chlorophytum (2), Clinogyne (1),Colocasia (4), Commelina (1), Crinum (1), Curculigo(1), Curcuma(12), Cyanotis (1),Cyperus(5), Dactylorhiza(1), Didymospermum (1), Dioscorea (36), Dracontium (1), Eleocharis (1), Eria (1), Etlingera (1), Eulophia (2), Gagea (1), Gastrochilus (1), Gonatanthus (1), Habenaria (3), Hedychium (2), Hypoxis (1), Kaempferia (3), Lasia (1), Leucojum (1),Lilium (1), Maranta (2), Paris (1), Phragmites (1), Phyllostachys (1),Plesmonium (1), Polygonatum (4), Potamogeton (5), Raphidophora (1), Remusatia (1), Sagittaria (3), Sauromatum (1), Scirpus (4), Scleria (1), Sinobambusa (1),Smilax (3), Theriophonum (1),Tigridia (1),Tulipa (2), Typha (5), Typhonium (3), Vanda (1), Yucca (1), Zingiber (4), Zizania (2).*

APPENDIX - II

Index to Families of Edible Roots and Underground Stems of Ethnic India

Name of Family	(Genus/Species)	Total Number of Species under a Genus
Alismataceae	(2/4)	*Alisma (1), Sagittaria (3).*
Amaryllidaceae	(3/3)	*Crinum (1), Hypoxis (1), Leucojum (1).*
Apiaceae	(12/21)	*Angelica (2), Bunium (1), Bupleurum (3), Carum (4), Chaerophyllum (2), Daucus (1), Ferula (3), Heracleum (1), Pleurospermum (1), Pastinaca (1), Peucedanum (1).* *Sium (1).*
Aponogetonaceae	(1/3)	*Aponogeton (3).*
Araceae	(14/31)	*Acorus (1), Alocasia (3), Amorphophallus (5), Arisaema (9), Colocasia (4), Didymospermum (1), Dracontium (1), Gonatanthus (1), Lasia (1), Plesmonium (1), Raphidophora (1),Remusatia (1), Sauromatum (1), Theriophonum (1).*
Araliaceae	(1/1)	*Acanthopanax (1).*
Arecaceae	(2/2)	*Borassus (1), Calamus (1).*
Asclepiadaceae	(10/17)	*Brachystelma (1), Carallumu (2), Ceropegia (6), Decalepis (1), Hemidesmus (1), Leptadaenia (2), Marsdenia (1), Pentatropis (1), Telosma (1), Utleria (1).*
Asteraceae	(8/10)	*Arcticum (1), Cichorium (1), Cirsium (2), Echinopsis (1), Elephantopus (1), Helianthus (1), Silybum (2), Taraxacum (1).*
Basellaceae	(2/2)	*Anredera (1), Boussingaultia (1).*
Boranginaceae	(2/2)	*Arnebia (1), Orthosiphon (1).*
Brassicaceae	(4/ 12)	*Armoracea (1), Brassica (7), Crambe (2), Raphanus (2).*
Cabombaceae	(1/1)	*Brasenia (1).*
Campanulaceae	(1/2)	*Codonopsis (2).*
Chenopodiaceae	(1/1)	*Beta (1).*
Commelinaceae	(3/3)	*Amichophacelus (1), Commelina (1), Cynotis (1).*
Convolvulaceae	(3/11)	*Erycibe (1), Ipomoea (7), Merremia (3).*
Costaceae	(1/1)	*Costus (1).*
Crassulaceae	(1/2)	*Sedum (2).*
Cucurbitaceae	(5/5)	*Kedrostis (1), Melothria (1), Momordica (1), Sechium (1), Solena (1).*
Cycadaceae	(1/1)	*Cycas (1).*

Name of Family	(Genus/Species)	Total Number of Species under a Genus
Cyperaceae	(4/11)	*Cyperus (5), Eleocharis (1), Scirpus (4), Scleria (1).*
Dioscoreaceae	(1/36)	*Dioscorea (36).*
Euphorbiaceae	(1/2)	*Manihot (2).*
Fumariceae	(1/1)	*Corydalis (1).*
Gentianaceae	(2/4)	*Nymphoides (3), Onosma (1).*
Geraniaceae	(1/1)	*Geranium (1).*
Hypoxidaceae	(1/1)	*Curculigo (1).*
Iridaceae	(1/1)	*Tigridia (1).*
Isoetaceae	(1/1)	*Isoetes (1).*
Juncaginaceae	(1/2)	*Triglochin (2).*
Lamiaceae	(5/8)	*Ajuga (1), Coleus (3), Dracocephalum (1), Marrubium (1), Stachys (2).*
Lauraceae	(3/3)	*Cinnamomum (1), Litsea (1), Persea (1),*
Lecythidaceae	(1/1)	*Careya (1).*
Liliaceae	(11/45)	*Allium (21), Asparagus (6), Asphodelus (3), Chlorophytum (2), Gagea (1), Lilium (1), Paris (1), Polygonatum (4), Smilax (3), Tulipa (2), Yucca (1).*
Malpighiaceae	(1/1)	*Aspidopterys (1).*
Malvaceae	(3/4)	*Abelmoschus (2), Abutilon (1), Althaea (1).*
Marattiaceae	(1/1)	*Angiopteris (1).*
Marantaceae	(2/3)	*Clinogyne (1), Maranta (2).*
Marsileaceae	(1/1)	*Marsilea (1).*
Menispermaceae	(1/1)	*Stephania (1).*
Moringaceae	(1/1)	*Moringa (1).*
Nephrolepidaceae	(1/3)	*Nephrolepis (3).*
Nyctaginaceae	(1/1)	*Mirabilis (1).*
Nymphaeaceae	(4/9)	*Euryale (1), Limnanthemum (1), Nelumbo (2), Nymphaea (5).*
Orchidaceae	(5/8)	*Dactylorhiza (1), Eria (1), Eulophia (2), Habenaria (3)), Vanda (1).*
Oxalidaceae	(1/1)	*Oxalis (1).*
Papaveraceae	(1/1)	*Dicentra (1).*
Papilionaceae	(10/22)	*Butea (1), Eriosema (2), Flemingia (3), Glycyrrhiza (1), Meconopsis (2), Mucuna (1), Pachyrhizus (2), Phaseolus (2), Pueraria (4), Vigna (4).*
Poaceae	(4/5)	*Phragmites (1), Phyllostachys (1), Sinobambusa (1), Zizania (2).*
Passifloriaceae	(1/1)	*Passiflora (1).*
Polygonaceae	(4/13)	*Antigonon (1), Polygonum (10), Rheum (1), Rumex (1).*

Name of Family	(Genus/Species)	Total Number of Species under a Genus
Polypodiaceae	(2/2)	*Blechnum (1), Pteridium (1).*
Portulaccaceae	(1/3)	*Portulaca (3).*
Potamogetonaceae	(1/5)	*Potamogeton (5).*
Ranunculaceae	(4/5)	*Aconitum (2), Clematis (1), Coptis (1), Paeonia (1).*
Rosaceae	(2/3)	*Cotoneaster (1), Potentilla (2).*
Sapindaceae	(1/1)	*Allophylus(1).*
Sauraceae	(1/1)	*Houttuynia (1).*
Solanaceae	(1/2)	*Solanum (2).*
Sterculiaceae	(2/3)	*Erythropsis (1), Sterculia (2).*
Taccaceae	(1/4)	*Tacca (4).*
Typhaceae	(2/8)	*Typha (5), Typhonium (3).*
Urticaceae	(2/2)	*Pouzolzia (1), Urtica (1).*
Vitaceae	(3/3)	*Ampelocissus (1), Leea (1), Vitis (1).*
Xyridaceae	(1/1)	*Xyris (1).*
Zingiberaceae	(8/30)	*Alpinia (6), Amomum (1), Curcuma (12), Etlingera (1), Gastrochilus (1), Hedychium (2), Kaempferia (3),Zingiber (4).*

APPENDIX - III

Index to Total Number of Edible Species of Roots and Underground Stems under Predominant Genera

Name of Genus	No. of Species	Name of Genus	No. of Species
Dioscorea	36	*Habenaria*	3
Allium	21	*Kaempferia*	3
Curcuma	12	*Merremia*	3
Polygonum	10	*Nephrolepis*	3
Arisaema	9	*Nymphoides*	3
Ipomoea	7	*Portulaca*	3
Brassica	7	*Sagittaria*	3
Alpinia	6	*Smilax*	3
Asparagus	6	*Typhonium*	3
Ceropegia	6	*Abelmoschus*	2
Amorphophallus	5	*Aconitum*	2
Cyperus	5	*Angelica*	2
Nymphaea	5	*Caralluma*	2
Potamogeton	5	*Chaerophyllum*	2
Typha	5	*Chlorophytum*	2
Carum	4	*Cirsium*	2
Colocasia	4	*Codonopsis*	2
Polygonatum	4	*Crambe*	2
Pueraria	4	*Phaseolus*	2
Scirpus	4	*Potentilla*	2
Tacca	4	*Raphanus*	2
Vigna	4	*Sedum*	2
Zingiber	4	*Silybum*	2
Alocasia	4	*Solanum*	2
Aponogeton	3	*Stachys*	2
Asphodelus	3	*Sterculia*	2
Bupleurum	3	*Triglochin*	2
Coleus	3	*Triglochin*	2
Ferula	3	*Tulipa*	2
Flemingia	3	*Zizania*	2

APPENDIX - IV

Index to Botanical Names of Edible Species of Roots, Tubers and Rhizomes of Ethnic India

Botanical Name	Page No.	Botanical Name	Page No.
Dioscorea melanophyrma	84	*Hemidesmus indicus*	104
Dioscorea nummularia	84	*Heracleum pinda*	105
Dioscorea oppositifolia	84	*Houttuynia cordata*	105
Dioscorea pentaphylla	85	*Hypoxis aurea*	105
Dioscorea prazeri	86	*Ipomoea aquatica*	106
Dioscorea puber	87	*Ipomoea batatus*	107
Dioscorea purpurea	87	*Ipomoea cairica*	107
Dioscorea quinata	87	*Ipomoea digitata*	108
Dioscorea rubella	88	*Ipomoea muricata*	108
Dioscorea sagittata	88	*Ipomoea paniculata*	108
Dioscorea sativa	88	*Ipomoea tuba*	109
Dioscorea tomentosa	89	*Isoetes coromandeliana*	109
Dioscorea triphylla	89	*Kaempferia angustifolia*	109
Dioscorea tuberosa	89	*Kaempferia galanga*	110
Dioscorea versicolor	90	*Kaempferia rotunda*	110
Dioscorea villosa	90	*Kedrostis rostrata*	111
Dioscorea wallichii	90	*Lasia spinosa*	111
Dracocephalum heterophyllum	91	*Leea edgeworthii*	112
Dracontium polyphyllum	91	*Leptadenia pyrotechnica*	112
Echinopsis nivensis	91	*Leptadenia raticulata*	112
Eleocharis dulsis	92	*Leucojum aestivum*	113
Elephantopus scaber	92	*Lilium nepalensis*	113
Eria andamani	93	*Limnanthemum indicum*	113
Eriosema chinense	93	*Litsea glutinosa*	114
Eriosema himaliacum	93	*Manihot esculenta*	114
Erycibe paniculata	94	*Manihot glaziovii*	115
Erythropsis pallens	94	*Maranta arundinacea*	115
Etlingera linguformis	94	*Maranta dichotoma*	116
Eulophia campestris	95	*Marrubium roylei*	117
Eulophia dabia	95	*Marsdenia roylei*	117
Euryale ferox	95	*Marsilea quadrifolia*	117
Ferula asafoetida	96	*Meconopsis aculeata*	118
Ferula galbaniflua	97	*Meconopsis paninculata*	118
Ferula narthrex	97	*Melothria heterophylla*	118
Flemingia strobilifera	97	*Merremia mammosa*	119
Flemingia tubersosa	98	*Merremia rhyncorhiza*	119
Flemingia vestita	98	*Merremia tuberosa*	119
Gagea persica	99	*Mirabilis jalapa*	120
Gastrochilus pandurata	99	*Momordica dioica*	120
Geranium wallichianum	99	*Moringa oleifera*	120
Glycyrrhiza glabra	99	*Mucuna atropurpurea*	121
Gonatanthus pumilus .	100	*Nelumbo nucifera*	121
Habenaria clavigera	101	*Nelumbo speciosum*	122
Habenaria commelinifolia	101	*Nephrolepis auriculata*	122
Habenaria susannae	101	*Nephrolepis biserrata*	123
Hedychium coronarium	102	*Nephrolepis cordifolia*	123
Hedychium spicatum	103	*Nymphaea alba*	123
Helianthus tuberosus	103	*Nymphaea lotus*	124

APPENDIX - V

Index to English Names of Edible Roots, Tubers and Rhizomes of Ethnic

APPENDIX - VI

Index to Hindi Names of Edible Roots, Tubers and Rhizomes of Ethnic India

Hindi Name	Page No.	Hindi Name	Page No.
Haldi	67	Kauraj	132
Halunda	168	Kesur	151
Hardi	67	Keu	61
Hathathjori	44	Khamalu	75
Hathchuk	103	Khantali	53
Hinduba	55	Khapparkaddu	52
Hing	97	Kharkar	133
Hing	97	Khate-chaval	140
Hingda	97	Khatta-mitha	128
Hingra	97	Kinda	105
Hukakanda	101	Kip	112
Hunraj Machoti	136	Kiralu	37
Jangli Pyaj	21	Kiri ki Kuri	38
Jangli Dhan	172	Kodan	83
Jangli Kla	120	Kohnda	145
Jangli Pindalu	100	Kooi	125
Jangli Shalgam	47	Kooka	125
Jangli –ushbah	155	Koorkan	58
Janglihaldi	65	Kudzu	144
Jethi-Madh	99	Kulanjan	27
Jikipota	128	Kumala	112
Kachalu	59	Kumali	112
Kachura	69	Kumbhi	50
Kaksa	120	Kumuda	126
Kala –zira	49, 51	Kusrunt	97
Kalazira	48	Lahsan	22
Kali musali	63	Lasan	22
Kali Sarson	46	Luduti	57
Kalihaldi	69	Madangameshvari	70
Kalim-sag	106	Mag	84
Kalizewar	49	Mahabari –bach	171
Kamal	121	Mahali	74
Kamal	124	Maida Lakri	114
Kangli	12	Makhana	95
Kanjan-Bura	109	Mameera	61
Kanphul	161	Man –alu	75
Kanta –Alu	84	Man –Alu	90
Kanval	121	Manghaii	84
Kanval	124	Mankanda	25
Kanval	125	Mankanda	25
Kanwal	121	Maslun	139
Kapurakachari	103	Meetha-dudia	135
Karchura	102	Mitha Alu	107
Karukandu	82	Motha	70
Kasaru	151	Motha	72
Kasni Kashini	55	Mothitrina	164
Kasturidana	12	Mou-alu	75
Kauathodi	131	Mudu-nirbisha	109

APPENDIX - VII

Index to Sanskrit Names of Edible Roots, Tubers and Rhizomes of Ethnic India

APPENDIX - VIII

Index to Regional Names of Edible Roots, Tubers and Rhizomes of Ethnic India